T0245322

AI IN CLINICAL PRACTICE

AI IN CLINICAL PRACTICE

A Guide to Artificial Intelligence and Digital Medicine

GIAMPAOLO COLLECCHIA
Primary Care Physician, Italy

RICCARDO DE GOBBI
Primary Care Physician, Italy

Academic Press is an imprint of Elsevier
125 London Wall, London EC2Y 5AS, United Kingdom
525 B Street, Suite 1650, San Diego, CA 92101, United States
50 Hampshire Street, 5th Floor, Cambridge, MA 02139, United States
The Boulevard, Langford Lane, Kidlington, Oxford OX5 1GB, United Kingdom

ISBN: 978-0-443-14054-9

For Information on all Academic Press publications
visit our website at https://www.elsevier.com/books-and-journals

Publisher: Stacy Masucci
Acquisitions Editor: Linda Versteeg-Buschman
Editorial Project Manager: Tracy I. Tufaga
Production Project Manager: Selvaraj Raviraj
Cover Designer: Christian J. Bilbow

Typeset by MPS Limited, Chennai, India

Dedication

As we write, the COVID-19 pandemic is underway, in which health professionals have demonstrated the uniquely human characteristics of responsibility, spirit of sacrifice, courage, and willingness to heal the suffering people.

To them, and in particular to the hundreds of colleagues who have sacrificed their lives or their health to fully fulfill their duty, we dedicate our book with great affection.

Contents

Foreword

Presentation

I start from what could be a conclusion: this book has a very particular importance, for its extremely updated cognitive contents, but even more for the method with which it asks not to "read" a very dense and legible text but to participate in a research project, looking for an answer to a question that has gone through all the many-few years (50 or so) from the time when technology (still with punch cards and calculators as large as rooms) began to cross paths with medicine:

> Is it possible, and where it leads, to adopt a mental language and a logic that aim to transfer the decision-making power of a discipline-culture such as medicine that has as its interlocutor the irreducible qualitative of the collective and individual life of people?.

The formulation of the question is certainly naive, but it reproduces literally the conclusions of the first-degree thesis of the Faculty of Medicine in Milan, supervisor GA Maccaro, on the role of information technology in medicine. It came back to my head, and I decided to risk the bias of self-citation (certainly unsuspected, given my present and historical digital illiteracy) because it seems to me to summarize well the second reason why this book is important: it is not a showcase of trendy novelties. It carries with it a whole story, with the questions of which it was necessary to interact: not about books, but within practices and roles that have seen a very large and conflicting evolution in the ways of producing and using knowledge.

Trial, epidemiology, data mining, big data, real-world evidence, decision-making, networking, algorithm, artificial intelligence, precision medicine, and so on up to the Nobel Prize given to economists in 2019 for having "discovered" that the evidence lies not in the models, but in the experimentation on the ground: these are not always the penultimate words of a technological and cultural market which is now among the most promising. They are identities and languages that carry within themselves all the promises, and at the same time, all the disenchanting ones, of a memory that, in order to be responsible, should have the minimum obligation of transparency and humility.

The following points correspond to those that a by now-old "informed user" has found treated in a very rare way in the book he wishes good luck:

1. The question of 50 years ago, naive or still with traces of "philosophy of science," meets, beyond a technological capacity light years later, a world context in which actors and interlocutors are others. In a global world and with structural proposals, medicine "speaks," "interests itself," and "describes" health, but no longer has fundamental decisions to make on health, which is a strategic area of economics, model development, choices of industrial policies, information and surveillance markets, and so on.
   This is the first decisive contribution of the book. Sober Robust. Factual. It gives the name to things and their owners. As a fact and a background: which can be forgotten, but which acts and is fundamental in the type and quality of the results. Like the "black holes" of algorithms, which obviously are perfectly known and programmed for/by those who produce them. An informed knowledge not so much on "complexity" (which is a fascinating term, which even seems to invite participation, and so on), but on the tangle-conflict of interests in which one moves, is the clearest "cultural gift and methodological" that comes to us.
2. Closely connected with the previous point is the context in which this book was born: a long history, individual and collective, very rich and original in general medicine. It turns out—discussed very well toward the end. I think it is good to propose a very brief reflection from now on. Beyond many contrary statements (which have become even ridiculous due to their formalism detached from concrete decisions in the COVID-19 era), the development of general medicine in the global context coincides, in fact, with its disappearance, in terms of cultural and institutional: if at least the problem of the identity of general medicine is taken seriously: it is a producer-interpreter of knowledge, as well as a guarantor of health-life rights (widespread and "near" to be universal), or is it a place of exhibition-propaganda-sale of products (very good or perfectly fake) of the micro-macro markets of winning welfare strategies (public or private)?Being at the forefront of the issues dealt with and discussed here from many points of view is a good test of identity and choice, just as it was at the time of the ancient and naive question, which was an indicator, as now, of a time of crisis. The direction had been important then, much hoped for, beyond expectations: to the point of translating in terms of laws into cultural junctions, of law, of dignity where society was crossing health care (194, 180, 833). In the current crisis, the demand that has become central in all areas, from privacy

to environmental sustainability, does not seem to have people's lives at the center, but rather their compliance with predefined "restrictions," waiting to know the direction and (unpredictable, even in times of AI, algorithms, and robots) ways out.

3. It is part of my now incurable biases (also because they were not felt as such, but always experienced as the only way to approach and understand realities in a nondependent way) to think that the most serious wish that can be made to those who open windows to the future and send messages, this can be translated into culture-research projects. The many exemplary scenarios that are proposed (it is one of the most stimulating features of the book) give the wish a sense of concreteness and realism. For general medicine, first of all. But not to look mainly within it, as if there was a need to formulate guidelines for a new sector of interest. To measure oneself with the many interlocutors mentioned in the previous point, starting from the daily life of all dis-encounters rather than from certainties. Entering the text as readers-users, but with the aim of leaving it as researchers: knowing very well (the text proves this repeatedly and with many concrete examples) that the priorities of the research are not those that are mainly or exclusively played in strictly areas clinical, or epidemiological, or evaluative of interventions, but those that cross and have among the "inclusion criteria" in research the variables of law, accessibility, dignity: not to apply the pass of a more or less qualitative research, but to deal with the structural biases in which all medical interventions live today, in terms of denominators or numerators of their populations and their samples. The *digital divide* that is clearly discussed in the book is not a variable that interests specifically "less developed" countries or human groups; it is instead one of the many dividing walls whose demolition is one of the important reminders, a true *take-home message*. A basic take home: not pizza delivered by black raiders who represent and summarize so well the basic flaw (the "plus," not the "bug") of the most successful algorithms, those of *just-in-time* distribution which has not even succumbed to the restrictions of COVID 19. Ken Loach from "Sorry, we missed you": mandatory.
4. One last note-recommendation-thanks. The bibliography references proposed gradually are impressive for their vastness of view and updating. They are essential to becoming coauthors of the book through research paths capable of moving in the long term; however, they are those that coincide with encounters with "people" (mythical or today's, poets of literature or life, and so on), which you are invited along the pages. Not only so much to enter more informed discussions (very well evoked and qualified on present or future bioethics, as well as on the legality or legitimacy of increasingly

intrusive apps), but to recognize that the sources that produce the evidence on life do not obey impact factor measures. The informatively flawless exploration of an area that overlooks and makes present imaginaries, gazes, visions of the world and things, needs (so says the subdued but constant memo of this book) to never leave, rather to feed on the imaginaries that are born, speak, decide, for better or for worse, starting from life. We all know. Is the story of a person who is the founding myth of the future told in this book summed up—by the deep fascination of sympathy-recognition-respect? Or for the esthetic choice of a lightness consistent with the needs of the market?—in a bitten apple: a symbolic "expulsion" of all expulsions from the freedom to choose seems relevant and central to the time—research that this book invites us to experience, to monitor, without an app, if and how the answer to the question from which we started is taking a direction that makes times of equality more accessible, even in medicine.

Gianni Tognoni
Epidemiologist and President of the ethics committee of the Milano Bicocca University

Acknowledgment

The authors would like to thank Elisa Rampin for the translation of this book.

Introduction

This book, the result of many years of study and continuous updating, intends to explain what artificial intelligence (AI) is, the maximum expression of digital medicine, its enormous potential but also the problems it raises, its limits, and also the risks of its large-scale use.

The first example of digital medicine with therapeutic purposes can be considered Eliza, a "Chatterbot" created in 1966 by Joseph Weizenbaum originally created for playful research, it showed unexpected therapeutic abilities; since then, progress has been discontinuous but overwhelming and in some respects uncontrolled.

Revolutionary discoveries have a common feature: they are regularly underestimated just as they introduce epochal changes without their contemporaries being aware of it. This happened in the past with the invention of printing, steam machines, industrial chemical processes, and also with the uncontrolled use of atomic energy. The same is happening, under our distracted eyes, for AI with its automation processes; everything is changing radically and irreversibly, and we not only do not know the potential of this process, but we ignore its limits and dangers.

The medicine of the 2000s is undergoing profound, radical transformations.

Recent advances in computational information technology have allowed the development of a new generation of systems capable of rivaling/surpassing human capabilities in certain domains or in specific tasks. These systems are also able to learn from their own experiences and take actions often not contemplated by designers. As stated by J. Kaplan, it is no longer true that "computers only do what they are programmed to do" [1].

Diagnostic systems based on AI, in particular on *deep learning*, and therapeutic-assistance systems based on avatars or robots are not only useful tools in the hands of doctors and researchers but are becoming the backbone of all advanced health systems, with the risk of reducing humans to mere cogs in a macro-system controlled by mysterious algorithms. In the words of Luciano Floridi, one of the most authoritative voices in the philosophy and ethics of information, they have become environmental, anthropological, social, and interpretative forces [2].

The digitalization process of medicine is advancing with overwhelming energy; even the austere and rigorous English NHS (National Health Service) has been influenced by it, and for some years since, it has been using the Babylon system as the first automated diagnostic level. AI is already able to make diagnosis and prognosis based on a simple radiological image or on a single photo of a histological preparation. In the psychotherapeutic field, avatars that compete with humans in the therapy of psychic problems are already functioning. The US Food and Drug Administration is actively facilitating the introduction of AI-based devices to the health market [3].

The digital high tide, the so-called "datanami," tsunami of data, is therefore submerging the millenary Hippocratic medicine in size and complexity without having provided evidence that this could mean real and lasting progress. The digital revolution is also changing the cultural, epistemological, and ethical paradigms of society. In this regard, Eric Topol [4], citing Joseph Schumpeter, a famous Austrian economist, speaks of "creative destruction" in reference to the transformation that generally accompanies great innovations.

The passage of eras coincides, for one of the unpredictable games of history, with the passing of the baton from the generations of doctors born in the postwar decade (the so-called "babyboomers") to the young 30 years old of today, professionally trained, motivated, sometimes fierce, but perhaps not aware that the overwhelming digital tide is pushing them towards new, unknown, exotic but insidious shores.

We believe that in the crucial moment in which the old "analog" generation, educated to critical and self-critical reflection, passes the baton to a new effervescent digital technological generation, this critical investigation of ours can take on the value of a testimony that refers to masters of thought that no algorithm will ever overcome. We mention only a few: Hippocrates for medicine but also and perhaps even more than our Father, Aristotle, Galileo, Descartes, Locke, Voltaire, Kant, Wittgenstein, and Popper, thinkers from whom generations of researchers have drawn ideas and methods for centuries.

If we constantly refer to these Masters, AI, which Tim Appenzeller has defined, in the journal *Science*, as "the apprentice scientist" [5], and digital medicine in general, can continue to be instruments of freedom rather than suffocating guardians of our thinking and of our actions.

For these reasons, we intend to offer practicing doctors and all interested people a short guide written by "not experts but diligent readers" for all those who consider themselves not experts but passionate and perhaps worried.

Advice to Readers

This book is structured on two levels: a synthetic one, aimed at rapid and intuitive learning, and a descriptive and analytical one, dedicated to those who want to reflect and deepen the polymorphic aspects of the digital-humanity interaction.

The tables at the beginning of each section and the brief summaries at the beginning of each chapter are intended for rapid reading, the individual chapters for analytical reading, and for a more detailed and in-depth reflection.

Given the continuous bursting evolution of research and applications, it is possible that some topics may need some integration in a short time: the last article we used for our book dates back to April 2020.

Enjoy the reading
Giampaolo Collecchia
Riccardo De Gobbi

References

[1] Kaplan J. Artificial intelligence. Guide to the near future. Rome: LUISS; 2017.
[2] Floridi L. The fourth revolution – How the infosphere is transforming the world. Milan: Raffaello Cortina Editore; 2017.
[3] Tirrell M. FDA moves to encourage AI in medicine, drug development. CNBC 018; https://www.cnbc.com/2018/04/26/fda-moves-to-encourage-ai-in-medicine-drug-development.html.
[4] Topol E. The creative destruction of medicine. How the Digital Revolution will create better health care. New York, NY: Basic Books; 2012.
[5] <https://science.sciencemag.org/content/357/6346/16>.

SECTION 1

Artificial intelligence: "What are we talking about"

CHAPTER

1

What does artificial intelligence mean?

Is the digital world a resource for our mind?

The great minds have already answered a large part of the questions we ask ourselves today

Plato narrates in Phaedrus that it was the God Theuth of ancient Egypt who gave writing to men; proud of the gift He asked King Thamus of Thebes why he was not grateful to him for this exceptional instrument ...

King Thamus replied, "O most ingenious Theuth, the creative power of new arts is one thing, it is quite another to judge what degree of harm and usefulness they possess for those who will use them. Writing will generate oblivion in the souls of those who will learn it: they will cease to exercise memory trusting the writing ... nor do you offer true wisdom to your pupils, but you only give the appearance of it because, thanks to you, they can have news of many things without teaching, they will believe they are very learned, while for the most part they will know nothing..." [1].

Plato thus expressed an atavistic fear that any technological advance could destroy what was painstakingly acquired, making men slaves to new tools, *a fear that various scholars express today amplified for the Internet.*

Many great minds have responded over the centuries to the enormous merits of writing books and annotations from books, among which we must at least remember *Seneca and Erasmus of Rotterdam* who, with different tones, style and topics, invited us to read, annotate, and rework what we, it was noted, and *Francis Bacon,* who advised noting reflections and sayings of illustrious men to draw inspiration for new ideas [2]. In more recent times *Umberto Eco* has provided a balanced and timely

continued

DOI: https://doi.org/10.1016/B978-0-443-14054-9.00019-3

(*cont'd*)

response on the digital world, and on the Internet in particular, considering them not threats but resources, as long as you approach them with a questioning, open mind, and a critical spirit [3,4].

We all have daily interactions with artificial intelligence (AI), which, since Alan Turing laid the foundations in 1936, has undergone a radical evolution in meaning and applications. It has become a contemporary tool that supports us every day in numerous activities (telephone assistants, web search engines, social networks, photo sharing, listening to music, spam filters, and commercial profiling). In fact, all these technologies are based on AI [5].

It is difficult to define it or explain its meaning. Francesca Rossi, Professor of computer science, presents it like this [6]:

Artificial intelligence is a scientific discipline that aims to define and develop programs or machines (software and/or hardware) that exhibit behavior that would be called intelligent if it were exhibited by a human being.

The goal of AI is, therefore, the creation of algorithms, robots, and technologies that use mathematics and statistics to express the complexity of human behavior in formulas in an "intelligent" way, apparently using reasoning, the ability to provide an opinion, forms of social behavior [7], and provide useful support for humans who seek to overcome their limitations and extend their capabilities.

The term intelligence also tends to confuse and to arouse both expectations and fears. In reality, it is already very difficult to define human intelligence, let alone how it works. We all have intuition, with which we try to judge a person by attributing to him a greater or lesser intelligence, but if we try to deepen the motivations, we only get a feeling of vagueness, of indeterminacy, and nevertheless its multiple definitions, starting from the one proposed by the Garzanti encyclopedia [8] and by Treccani [9].

For our purposes, it may suffice to say that machines are not intelligent in the usual sense. They cannot "think like humans, only faster and more efficiently." Tools available today have nothing remotely resembling the intelligence of a mouse or a dog. Yet they have a tremendous ability to successfully solve problems and to pursue objectives set by human beings.

In short, we can agree with the scientist Francesca Rossi that human and artificial intelligence "share" a mode of cognitive activity, described by psychologist and Nobel laureate for Economics Daniel Kahneman, which divides our mental activities into two categories: slow thinking and fast thinking. Fast thinking (system 1), consisting of immediate decisions based on intuitions and emotions, is activated when an almost automatic response is needed. Slow thinking (system 2) instead requires reflective, analytical reasoning. The two ways of thinking are not dichotomous but interact and sometimes conflict.

Even for AI, according to F. Rossi, despite the fundamental differences between men and machines, we can distinguish slow thinking and fast thinking: "The AI techniques that allow you to solve problems in an always correct and optimal way and follow a logical and symbolic reasoning, are the basis of slow thinking of AI, while those that allow you to solve even problems defined in a vague way, but do not ensure complete correctness, are examples of fast thinking of AI. It is up to us to decide how to make these two ways of "thinking" interact in AI."

We could consider the simple algorithms used by the so-called "expert systems," based on IF... THEN, to the slow thinking, the more complex machine learning to the slow one.

In reality, machines and people are enormously different. Machines have, in fact, a practically infinite memory and enormous information processing capacity so that no natural intelligence can compete with AI in terms of speed and computational capacity. The human brain, at the moment, has inimitable perceptive abilities (the "Gestalt perception" well described by psychologists), greater capacity for reasoning and abstraction, greater critical and self-critical abilities, and unsurpassed creative abilities.

References

[1] Plato. Works. vol. I. Laterza Bari Ed.; 1967.
[2] Quoted in Carr Nicholas. Does the Internet make us dumb? Raffaello Cortina. Milan; 2013.
[3] Eco U. Apocalyptic and Integrated. Milan: Bompiani Ed.; 2001.
[4] Eco U. <http://www.italynet.com/columbia/ecopage.htm>.
[5] The term Artificial Intelligence was used for the first time in 1956 at a summer workshop at Dartmouth College in Hanover (New Hampshire, USA), by John McCarthy, a young American computer scientist, pioneer and inventor, who defined it as "science and 'engineering to make intelligent machines."
[6] Rossi F. The border of the future. Milan: Feltrinelli; 2019.
[7] <https://www.ubs.com/microsites/artificial-intelligence/en/new-dawn.html>.
[8] <http://www.garzantilinguistica.it>.
[9] <http://www.treccani.it>.

CHAPTER

2

How does Artificial Intelligence work?

Summary

The term "Artificial Intelligence" was born and spread to explain with an analogy that "digital" can carry out complex and innovative projects.

In reality, AI has nothing to do with the intelligence of humans: it is a real "New World" that we must at least partially know in order not to be overwhelmed.

In this chapter, we present the foundations of this gigantic building by sketching, schematically but rigorously, its essential characteristics and underlining its important differences with respect to the human mind, which still remains the most complex and wonderful object in the universe.

The *anthropomorphist orientation*

According to the philosopher Eric Sad [1], man has always been animated by the passion of generating artificial doubles of himself, endowed with kinesthetic, sensory-motor, proprioceptive, and above all, cognitive qualities. The real aspiration would be to create entities much more powerful than us, starting from our constitution, capable of establishing an organization that is more reliable and perfect than reality.

In the context of AI, the ambition is to build systems modeled on the body, and in particular, on the human brain. Proof of this is the "neuro lexicon," which represents the backbone of research in computational science laboratories; synaptic chips, genetic algorithms, and neural networks.

The tendency to accord a preeminent position to neuroscience means that various disciplines add the prefix "neuro" to their name; neuropolitics, neuromanagement, and neuroeducation.

DOI: https://doi.org/10.1016/B978-0-443-14054-9.00008-9

The fundamentals

The discipline of "Artificial Intelligence" was born in Dartmouth (USA) with the "Summer Research Project on Artificial Intelligence" in 1956. The first stated goal was to reproduce human intelligence and the first methodological approach was that of analogy. (1)

Even then AI was, however, first of all digital (from digit and from the Latin digitum, finger**);** digital because each elementary component of computers can only assume finite states, in particular two, 0 or 1 (on or off, true or false), which allow representing the smallest unit of information, the bit (contraction of *binary digit*, which can take the values 0 or 1) [2]. In this way, the AI can process, once encoded in the binary sequence of 0 and 1, information of any type (numbers, words, images, sounds, or other physical entities), **through the use of appropriate algorithms, increasingly complex, which can be translated into mathematical terms by means of the "Godel numbers**."

In the past, data processing had **always been** analog; the electrical signals were not binary, discontinuous, but continuous. They could, therefore, assume the infinite range of values of the quantities they reproduced [3].

The experts had, therefore, trained themselves in an analog world and considered digital language only a useful tool and sometimes an annoying task: until 1984, the American Association for Artificial Intelligence (AAAI) rejected innovative hypotheses based on the systematic and sequential use of statistical evaluations and confirmed the traditional static approach. (1)

The purely analog-imitative approach has, however, not allowed substantial progress for decades, as long as a group of younger scholars and in particular Ray Solomonoff, thanks to a masterful and intensive application of Bayes' theorem (test-error-retest); they managed to make the decisive leap in quality; they passed from a purely deductive methodology with a low innovative level to a prevailing inductive method with a high innovative capacity, even if forced to continuous checks and retests, which was, however, made easier by the great speed of the machines.

It was "the egg of Columbus." It is, in fact, the procedure used by living beings for billions of years and that the current very powerful computers can handle in a few weeks.

Machine learning

In practice, the systems based on Machine Learning are "trained" through the presentation of huge datasets consisting of millions of digitized images (for example, radiographs, photographs, electrocardiograms), already classified on the basis of a *gold standard* (usually a defined diagnosis by a majority by a group of specialists), which can, however, be multiple and in any case subject to the risk of uncertainty in interpretation. Depending on the choice of diagnosis to use for algorithm training, the levels of accuracy can be different. In some studies on the diagnostics of diabetic retinopathy, the use of optical phase coherence tomography (OCT) as a *gold standard*, instead of the majority decision of a group of ophthalmologists, according to some researchers, could have increased the high levels of accuracy obtained [4].

After this period of "*supervised learning*," a phase follows in which new images are presented to the model, always ordered by the experts, but without the system being shown the "correct" classification. Its predictive capacity and autonomous diagnostic accuracy are, therefore, observed with respect to cases already correctly classified. This process can be repeated until very high predictive accuracy levels are achieved.

Deep learning-neural networks

Although computer-aided diagnostics is nothing new, the profound change lies in the development of deep learning (DL), based on Neural Networks, generally consisting of several successive layers, each composed of circuits that interact with each other inside of each layer; the first activated layer transmits to the second layer a solution, which is tested and transmitted to the third layer. The process is repeated several times in several layers, and the results provided are the result of random combinations of data that are repeatedly filtered by eliminating the combinations deemed useless, impossible, or inconsistent until reaching an optimal result (see glossary).

In any case, the approach is probabilistic (i.e., the machine associates a certain probability with the fact that there is the object to be examined in the image it received). Furthermore, the system always has an error rate, even if it is small. On the other hand, in the supervised approach, if the data entered is accurate and the rules cover all possible cases, the resolution of the problem is safe. The two approaches are often integrated.

The black box model: AI is like the brain: you cannot cut off your head and see how it works

This sentence, by Andy Rubin, cofounder of Android [5], effectively summarizes the most important and embarrassing data for AI experts, who are often unable to clarify how the system "reasons" and how it comes to offer us those type of solutions (which generally work for another...). This is why the procedures are defined as "opaque but efficient."

In practice, when a deep learning model predicts the indication for a biopsy investigation of a skin lesion with a high probability, it is a melanoma (it is a melanoma with a probability of 0.8), no one can establish the basis on which characteristics of the lesion the machine has elaborated this prediction, so much so, that the operating mode of these systems has been defined as a black box model [6].

Nello Cristianini, a great Italian AI expert, currently in Bristol, expressed the situation well by defining the new AI as an alien genius: we do not know how it works, and we have given up understanding why, but there is no doubt that it works in many sectors [7].

According to Eric Topol, the "black box" aspect of AI would also be emphasized due to the excessive expectations towards AI. Algorithms are not infallible and are not transparent in their computational steps, but many aspects of clinical practice are also unclear, sometimes inexplicable, for example, the prescription of therapies whose mechanism of action is not completely known but are generally more tolerated as "human."

Alessandro Vespignani, Professor of Physics and Computer Science in Boston and world expert in the field of scientific predictions and network theory, believes that machine learning algorithms, and in particular neural networks, allow computers to acquire a type of implicit knowledge, even if they are unable to explain the reasons of their results. This would allow them to bypass the so-called "Polanyi paradox," according to which "we know more than we can explain." This sentence represents the fact that we know a lot about how the world works but are unable to make this knowledge explicit, which Polanyi called "implicit." Hence an important question: Do algorithms really produce knowledge if they do not actually deepen our understanding of how the world works? The empirical answer of researchers and users of algorithms, especially in the industrial world, is that "we are paid to work, not to understand why and how" [8]. (see Cabitza table)

One approach to soliciting physicians to accept this "inscrutable" system could be to use what the learning software produces to train another, more transparent model that provides analyzable and understandable human responses [9].

The application of AI in real contexts can determine numerous potential advantages, such as speed of execution, potentially reduced costs, both direct and indirect, better diagnostic accuracy, greater clinical and operational efficiency ("algorithms do not sleep"), the possibility of access to the assessments even for people who cannot otherwise benefit from them for geographical, political and economic reasons. To have a reference scale, remember that a medium-power computer has a processing capacity that is at least 1 million times faster than the human mind: translated into numbers, this means that these common computers in an hour can process the same amount of data that an expert technician processes in over 4 years of work at 40 hours per week (without holidays!).

The Joker of statistics: Bayes

The statistical approach used in the treatment of these enormous quantities of data is usually of the Bayesian type[1]: the first results are generally imprecise and constitute the "a priori probabilities" that are processed by formulating less imprecise values of "conditional probabilities": these time are treated as a priori probabilities of further probabilistic estimates in an increasingly accurate and refined process.

The use of the enormous computational speed of computers to make probabilistic estimates based on the Bayes Theorem has opened new perspectives for AI. The results are there for all to see, and in various respects, excellent: one of the main practical applications of machine learning (ML) in medicine is, for example, the interpretation of clinical, radiological, histological, and dermatological data, more accurately and quickly than with the classical methodology. Human biology is, in fact, so complex and the expansion of knowledge so rapid that no natural intelligence can compete with AI in terms of speed and information processing capacity.

The decision support systems based on the DL have proved to be valid in various areas, in particular in the diagnostics of diabetic retinopathy and macular edema and skin tumors (see chapter...), reporting a level of accuracy equal to that of expert specialists: some ML systems applied to radiological diagnostics can now compete with the best radiologists [10].

Other possible applications are numerous; for example, the algorithms are able to systematically analyze electrocardiographic traces of many days and identify minimal variations apparently related to the risk of sudden death.

But the most fascinating and disturbing implications have emerged from other applications for diagnostics and therapy, so much so that the production and use of software, therapeutic video games, virtual reality platforms, avatars, and robots in the healthcare sector has now almost completely escaped from doctors and researchers' control and is largely designed and managed by IT experts and placed on the market according to methods and criteria chosen by the giants that operate on the Internet: for several years IBM with Watson, Google with Deep Mind [11] and recently the Babylon system [12] offer AI software that

1 Bayes' theorem makes it possible to estimate conditional probabilities with increasingly accurate levels of precision, that is, the probabilities that an event B will occur following the occurrence of a known event A: on the basis of the calculation, the most probable solution is chosen, often you test it in virtual environments and correct it one or more times.

formulates diagnoses and proposes therapies, without significant reactions from the international medical class.

Evolutionary adaptation software and genetic algorithms

They are based on this versatile theorem and reproduce in the research process; on the one hand, random genetic mutations, which are sometimes very useful, and on the other hand, the mechanisms of natural selection; in other words, many different solutions are tried, which are gradually eliminated by the software's selection mechanisms; if we remember that modern, powerful microprocessors carry out thousands of calculations per minute, we understand that computers are not intelligent, but they are only formidable workers who sooner or later "*get us* right" by mathematical laws [13].

Neural networks and deep learning in reading images

To understand, albeit superficially, the functioning of AI systems, we briefly describe how "neural networks" are able to "read" radiological images. *For this it is necessary to dwell on the radical differences in the acquisition and interpretation of images between humans and the neural networks of deep learning* (see also the glossary).

Like humans do

> "If the human brain were so simple that we could understand it, we would be so simple that we could not understand it" ***(GE Pugh, 1977).***

The recognition of images by humans follows complex and not yet fully known mechanisms: in summary, however, we can, first of all, distinguish a perception of the stimuli by the eyes and the transmission to the nerve centers that organize the perceived entities in various ways: figure- background, proximity-distance, continuity-discontinuity, similarity-diversity, natural "good shapes" and unnatural forms, closing and opening of lines, surfaces, and spaces.

Thus we arrive at an overview that our nerve centers will catalog according to models and maps we have built during our life. The process is fast and largely unaware and allows, for example, in medicine the "diagnosis at first sight" (spot diagnosis).

As the perfect digital cameras do

The digital image is made up of optical points (pixels), each of which is associated with numerical coordinates: the computer, based on the coordinates that refer to the position and function of the point (e.g., central or peripheral point, belonging to the figure, to the outline or the background, etc.) assigns to each pixel a position and a brightness that will contribute, with those of the other pixels, to form the final image.

In other words, the algorithms that receive the data relating to the pixels are able to extract various information from them that will allow the image to be reformed according to various levels of precision and accuracy.

There are many systems capable of recognizing images based on low-medium level indexes, while algorithms that are able to understand the content of an image in particular contexts are still being researched: for these reasons, computers have a very high sensitivity and precision in interpreting images of physical-mathematical phenomena while they have much greater difficulty in interpreting phenomena such as facial expressions.

The complexity of the models is determined by the complexity of the problems. Almost 130,000 images were required to classify diabetic retinopathy and skin cancers. The application of DL techniques to clinical conditions, such as heart or renal failure, would require tens of millions of examples, as well as in the case of data from different sources, such as texts, laboratory tests, and vital signs, which are difficult to obtain and use [14].

"The best approximation to what we know is that we know next to nothing about how neural networks actually work and what a really profound theory would be like."

(Boris Hanin, Texas A&M University mathematician and *visiting scientist* at Facebook AI Research).

Neural networks are inspired by the human brain and its ability to function by accumulating ever greater abstractions. The complexity of thinking, in this perspective, is then measured by the range of minor abstractions that can be drawn from and by the number of times it is possible to combine lower level abstractions into higher level abstractions, such as when humans learn to distinguish the dogs from the birds. Moreover, if abstraction is natural for the human brain, for neural networks, it involves complicated processing.

Neural networks are made up of structures of basic computing units called "neurons," interconnected in various ways and in more layers (the greater their number, the "deeper" the network). The first layer accepts input (for example, the pixels of an image), the last one produces the output (a description of the image content), and the intermediate, or "hidden," levels generate arithmetic combinations of the input.

continued

(*cont'd*)

Crucially, neural connections are not fixed in advance but adapt in a trial and error process. For example, a neural network that has the task of recognizing objects in images is presented with images labeled "dog" or "cat"; for each image, the system tries to guess the label; when it makes a mistake, with a simple calculation exercise the strength of the connections that contributed to the wrong result is changed, in an adjustment of the "weights" that amplify or attenuate the signals carried by each connection.

In fact, if the answer is wrong, an algorithm called *backpropagation* repeats the process by adjusting the intensity of the connections, to obtain a better result next time.

Starting from scratch, when the network does not know what an image is, much less an animal, the results are no better than what would be obtained with the flip of a coin. But after thousands of examples, the system reaches the level of a human being.

The idea is that each level combines different aspects of the previous level. "A circle is a set of curves in many different places, a curve is a set of lines in many different places," says David Rolnick, a mathematician at the University of Pennsylvania.

For imaging activities, designers typically use "convolutional" neural networks, which have the same pattern of connections between layers repeated over and over.

In general, we rely largely on experimental tests: we spin 1000 different neural networks and observe which one does the job. "In practice, these choices are often made by trial and error," says Hanin. "It's a bit of a complex approach because there are endless choices and we don't really know which one is best."

http://www.lescienze.it, 9 February 2019

References

[1] Pareschi R, Dalla Palma S. Intelligenza Artificiale, Milano: Hachette, 2019.
[2] Henin S. AI Artificial intelligence between nightmare and dream. Milan: Hoepli; 2019.
[3] Fabris A. Ethics for information and communication technologies. Rome: Carocci Editore; 2018.
[4] Wong TY, et al. Artificial intelligence with deep learning technology looks into diabetic retinopathy screening. JAMA 2016;316:2366–7.
[5] Deluzarche C. Deep learning, le grand trou noir de l'intelligence artificielle. Maddyness; 2017. https://www.maddyness.com/2019/08/20/ia-deep-learning-trou-noir-intelligence-artificielle/.
[6] Salovey P, Mayer J. Emotional intelligence. Imagination Cogn Personal 1990. Available from: https://doi.org/10.2190/DUGG-P24E-52WK-6CDG.

[7] Cristianini Nello. New scientis machines that think. Bari: Dedalo Editions; 2018.
[8] Vespignani A, with Rijtano R. The algorithm and the oracle. Milan: Il Saggiatore; 2019.
[9] New Scientist. Machines that think. The new era of artificial intelligence. Bari: DEDALO Editions; 2018.
[10] Maddox TM, Rumsfeld JS, Payne PR. Questions for Artificial Intelligence in health care. JAMA. 2019;321(1):31–32. Available from: https://doi.org/10.1001/jama.2018.18932.
[11] https://deepmind.com/.
[12] https://www.babylonhealth.com/.
[13] Kurzweil A. How to create a mind. The secrets of human thought. Milan: Apogeo Ed; 2013.
[14] Wang F, et al. Deep learning in medicine - promise, progress, and challenges. JAMA 2019;179:293–4.
[15] https://www.lescienze.it/news/2018/10/03/news/appearning_automatico_elefante_stanza-4138307/.

CHAPTER

3

The limits of algorithms

For several decades, cognitive psychologists who deal with the correctness of logical inferences have identified a bias (a basic reasoning error) particularly frequent in the digital world, the "pro-innovation bias," that is, the enthusiastic acceptance of apparently innovative achievements, as "original," would certainly be useful and effective.

In reality, centuries of research and experimentation in all scientific disciplines lead us to be cautious: any new acquisition should certainly not be rejected at first but should be evaluated in all its potential and inserted in the real context of the application, evaluating its potential negative effects.

This chapter is dedicated to researching the limits of some essential tools for the entire digital world.

"Data accumulation is no more science than a pile of bricks is a house" (Henry Poincaré, physicist and mathematician).

The overwhelming success of Big Data processing algorithms has led various experts (united by great hyper-specialized expertise but limited knowledge of the history and philosophy of science) to enthusiastic statements regularly denied by more rigorous analyzes of reality.

C. Anderson, former editor of Wired magazine and digital theorist, for example, even spoke of the reversal of the traditional learning pyramid. *The recent availability of large amounts of data, coupled with statistical tools to manage these numbers, offers a whole new way to understand the world.* ***Correlation replaces causality, and science can advance even without coherent models, unified theories or mechanistic explanations***...

According to this current of thought, algorithms deliberately overturn the classical scientific method (hypothesis, model, experimentation), because the models emerge directly from the data, from the associations *correlation is enough*, they are not guided by hypotheses. When the data is sufficient, the numbers speak for themselves. Statistical algorithms, inserted into the largest processing clusters ever, can identify usable patterns and schemes, without science knowing how to explain their origin or put them in context.

DOI: https://doi.org/10.1016/B978-0-443-14054-9.00011-9

This large group of experts is certainly a resource for the community, but each of their statements should be evaluated with the same methodological approach that allowed the emergence of their intuitions: in a nutshell, it is the one used by Albert Einstein who, after having formulated the theory that would change the world took care to identify an event that could confirm or falsify it: the famous phenomenon of the curvature of light, verified by Eddington in the eclipse of 1919.

In our limited possibilities, we will try to use this approach, critical and constructive rather than triumphalistic and apodictic, to identify the notable advantages but also the insidious limitations of the Big Data analysis algorithms, recalling a fundamental axiom of probability mathematics: the more numerous are the data and interactions between the entities involved the more often random, and insignificant correlations are found.

It is not the technologies that are decisive but the ability to extract value from their use

As stated by Federico Cabitza, *"Technology should make us see further and deeper, but leave the interpretation of what we see to us."* The data itself is, in fact, useless. To paraphrase the Nobel laureate in economics Ronald Case, *"if you torture the data long enough, they'll confess anything."*

The data is not a closed, "given" entity but a social construct, the concrete result of specific cultural, social, technical, and economic choices made by individuals, institutions, or companies to collect, analyze and use information and knowledge.

The very concept of *raw data* is an oxymoron: data that is not contaminated by theory, analysis, or context does not exist but is always the result of operations and elaborations of various kinds. In order for them to be truly useful, they need to be selected, structured, and interpreted [1].

Contrary to the claims of C. Anderson, the enormous amount of data requires, even more than in the past, an enormous interpretative effort, which computers are not (for now?) Able to perform independently.

Another limit to the reliability of the data is linked to uncertainty, an inevitable variable in medicine, a discipline characterized by large gray areas of knowledge due to the incomplete domain of available knowledge and the intrinsic limits of medical knowledge.

Multinationals such as Google, IBM, Microsoft, and Facebook, universities and research centers, public and private, national and international, are strongly present in the sector, attracted by the potential of these technologies. Moreover, as stated by E. Santoro, Director of the Laboratory of Medical Informatics of the Mario Negri Institute in Milan, digital health is also studded with sensational failures. Typical is the case of Google's blocking of the "smart" contact lens project to measure glucose levels, which began in 2014. The reasons are due to the lack of data able to demonstrate a real correlation between the glucose present in the tears and the concentration in the blood, and therefore to the lack of valid scientific support to accompany the request for registration as a medical device.

The lesson is that large investments may not be enough. More scientific research is needed on several levels, in particular, studies to verify the reliability of the hypotheses with realistically achievable and verifiable objectives. Above all, the announcements of the IT big names and the so-called "medical futurists" (the category of doctors and technologists who see solutions to many problems in today's projects) should not be overemphasized, too often brought, in the conjugation of verbs, to the conditional and to the future, preferring instead news accompanied by results, the result of clinical studies that demonstrate its real reliability and effectiveness [2].

On the contrary, as Santoro recalls, as early as 2015, articles were published claiming that by "using smart contact lenses, it would be possible to monitor, through the tear fluid, the glucose level in patients with diabetes."

Randomized clinical trials (RCTs), the methodologically most valid studies in the context of AI, are rare, if not entirely absent. The same tools approved by the US FDA, a software for diagnostics of diabetic retinopathy and recently an alert system for the presence of stroke at CT scan diagnostics [3], have not been validated by clinical trials. If it is true that even traditional diagnostic tools are not always subjected to rigorous studies, their use is of simple support for doctors, who have the final decision. In the case of AI, the system provides both information and operational advice [4], **with potentially dangerous consequences for patients**.

Recently, a clinical trial has finally been created, ironically called HYPE ("needle, puncture" but also "advertising message") (Hypotension Prediction During Surgery), on a decision support system based on machine learning. The study randomized 68 patients undergoing elective noncardiac surgery into two groups, one control with standard intraoperative treatment and another with AI-based alarm-guided intervention [5]. The aim was to evaluate the reduction in the intensity and duration of intraoperative hypotension, and in fact, the group that used the AI performed better. This is not the place to enter into the merits of the results and limitations of the study, for example, being only a pilot study, a sort of phase 2b of drug trials, on a small series, a highly controlled setting. Instead, we would like to point out the importance of starting to carry out trials as an indispensable tool for evaluating the clinical effectiveness and efficiency of AI systems. In particular, as in the case in question, by comparing teams of doctors who use AI systems and teams that do not while avoiding less significant comparisons between doctors and AI.

Possible criticalities related to the use of artificial intelligence systems in medicine [6]

Black-boxing: inscrutability of algorithms, defined as "oracular," in the sense of accurate but not associated with explicit and meaningful explanations of the answers they process and provide.

Overreliance: the risk that doctors may develop an unjustified and excessive reliance on automation capabilities, regardless of the variability and uncertainty of the context.

Overdependence: real dependence on these systems (overreliance).

Deskilling: reduction of the level of competence required to perform a function (de-qualification), when all or some of the components of the tasks have been automated.

Context-underevaluation: giving greater importance to data, which can be easily expressed and codified in numbers, with respect to the context, difficult to represent and explain (desensitization toward the clinical context).

Epistemic sclerosis: the vicious circle in which the *patterns* to which artificial intelligence systems are sensitive, considered ontologically reliable, are highlighted and suggested to doctors, who become less sensitive to identifying others, or the same ones, but independently.

The limits of predictive models

The predictive capabilities of AI in the prognostic field are particularly attractive and able to stimulate investments for millions of dollars by industries. The impact of AI on the entire global economy is estimated to reach 14% by 2030 [7]. After the hopes created in the 60 s by the so-called "prognostograms," in the following 40 years, the studies were amplified and the techniques enhanced, but the analytical models in the cardiovascular field have not achieved great success in clinical practice for a variety of reasons, especially for the greater confidence on the part of doctors in their intuition than in the results of the algorithms, also due to the lack of training in probabilistic estimates [1].

Advanced AI technologies currently have the potential to solve one of the most difficult problems in the medical field; controlling complex systems, understanding them, and using them to make predictions. The processing of patient data by ML algorithms could effectively build effective predictive models to improve clinical performance and reduce healthcare management costs.

The high expectations toward AI and the propensity toward the uncritical acceptance of innovation, such as the so-called "pro-innovation bias," however, risk underestimating the risks related to an acceptance of related technologies not motivated by certain evidence. The studies are heterogeneous, observational, and retrospective, not validated in the clinical setting.

The data necessary for the training of ML algorithms to process diagnostic processes in predictive models are often of suboptimal quality because they are not subjected to that process of "cleaning" and reprocessing (stratification by cohorts, filtering by quality level, etc.), which would be unsustainable in daily clinical practice and therefore may not be able to provide implementable answers for clinical decisions and treatments, also because, at times, they could "learn" the mistakes of "natural" intelligence, that is, of professionals who provide, *training data* "from training" for algorithms.

Methodologically robust, prospective works are indispensable, comparing the AI models and the usual practice on clinical outcomes, published in journals subjected to peer review. Only in this case will AI systems effectively applicable in real rather than experimental clinical contexts be created in the near future.

Many other data are not organized and standardizable: electronic health records are, for example, heterogeneous and unstructured. The information often concerns general pathological conditions, for example, diabetes, but not more specific aspects, such as duration, severity, and pathophysiological mechanisms. The same systems of *natural language processing* (see glossary) can analyze unstructured data, for example, clinical notes, but they are still raw and not able to elaborate all the possible nuances. Electronic records, especially hospital records, also tend to overestimate the sickest populations compared to the healthiest ones, which have fewer opportunities to be included in the archives [8].

In addition to being unstructured, the so-called "real world" data are not always available, they are not everywhere, and their value is also limited to the collection setting. In addition, many data are lacking, for example, previous conditions or diseases that the patient does not remember or that he has lived in remote times or in places other than current ones. For many important syndromes, such as heart failure, there is no consensus on the standard criteria on which to train the algorithms [9].

The use of big data, therefore, has limits that depend on their very nature, as data and not values: risks of bias in the selection of the sample, in the collection and interpretation of the information that is processed, capable of threatening the validity and the generalizability of the conclusions. For example, spurious correlations can be revealed; the more possible the larger the population under study.

Furthermore, the contextual aspects, difficult to explain in quantitative terms, can be underestimated and underrepresented; just think of the conditions that cannot be clearly defined in terms of pathology, the "fragility," the conditions of extra-clinical discomfort, the psychological factors, social, family, conditions of economic or cultural disadvantage, the type of health systems, which always affect the epidemiology of diseases and the availability of treatments, and therefore, on the clinical management of the patient.

Do algorithms photograph reality?

Algorithms also tend to learn and thus perpetuate social reality, including the prejudices and preconceptions that characterize it, especially of a racial and gender type. For example, the underdiagnosis of myocardial infarction is described in elderly women, who often present with atypical symptoms. Or the reduced risk assessment of breast cancer with genetic tests in black women, usually subjected less frequently to the search for mutations [10].

Sometimes for the training of the algorithms, particular cases, rare, not representative of the normal case series, can be presented, realizing a sample selection bias. The case history also changes with the passage of time, both in the sense of improvement and worsening of diseases, and this can lead to a mismatch between training and operational data of decision-making output.

In general, models trained for ONE problem, such as diagnosing pneumonia, may lose effectiveness in diagnosing incidental findings, such as suspicious images for neoplasia. The role of the human is, therefore, still of great value in the age of the elderly population, characterized by multimorbidity [11].

It is, therefore, essential to carefully monitor decision-making systems through adequate funding, evaluate their performance, and update the input data according to the evolution of scientific knowledge.

Unfortunately, we must also foresee the possibility that the algorithms are programmed to achieve higher sales for some drugs, laboratory tests, or devices, without the users obviously being aware of the manipulations.

In a famous study by Esteva et al., [12] both humans and algorithms have found it difficult to differentiate benign and malignant melanocytic lesions, but dermatologists have erred on the "side of caution," even risking overdiagnosis, while this attitude was not present in the machine learning system. **In practical application, on the other hand, an evaluation of overdiagnoses (false positives) and missed diagnoses (false negatives) should be envisaged** [13].

Wearable devices can also allow early diagnosis, for example, in Parkinson's[1] disease or Alzheimer's disease, but currently, the real clinical benefits of a diagnosis in the very early stages of the disease are not highlighted. Sometimes risk reduction has benefits that are valid at the population level but minimal in the individual. This reduces the ambition to pursue great goals in favor of realistic and personalized expectations.

As with the human brain, it is also difficult for AI models to recognize patterns for diagnosis and treatment. In particular, data from patients with low socioeconomic status may be unreliable due to a variety of reasons, including the fragmentation of care in different healthcare institutions. In these cases, the algorithms can be less accurate and increase health inequalities [14]. For example, one study found that clinical data alone have limited predictive power in the case of patients whose risk of rehospitalization is mainly dependent on social determinants[1].

Caruana et al. [15] found a clinical situation in which the predictive effectiveness of decision support systems was technically valid, but in practice, misleading. In over 14,000 pneumonia patients, different algorithms have been evaluated to predict the risk of mortality. The result was that patients with a history of asthma were classified as at lower risk of death than nonasthmatics[1]. The unexpected result was explained by the fact that patients with pneumonia and a history of asthma were generally admitted to the ICU, and the lower mortality was likely due to a tendency for doctors to treat them earlier and more aggressively. In practice, it is confirmed that formally perfect algorithms can be wrong due to the incompleteness and variability of the data entered.

An example of bias in data interpretation is the so-called "Texas sharpshooter fallacy[1]", a mistake that is made when differences in data are ignored and similarities are accentuated at the same time, thus arriving at an incorrect conclusion. This is related to the *clustering*

continued

1 The name comes from a joke about a Texan who would have fired a few gunshots on the side of a stable and then painted a target centered on the group of closest shots, and thus, be able to claim to be a sharpshooter.

(*cont'd*)

illusion, a human cognitive tendency to interpret the existence of patterns or patterns even where they do not exist. The Texas Sniper Fallacy often occurs when you have a large amount of data available but focus on only a small subset [16].

A similar mechanism, as regards the algorithms, occurs in the so-called "overfitting," that is, when the algorithm fits too well (over) to the training data but loses in generality; in practice, it has too many parameters relating to the number of observations, a condition that leads to overestimating its usefulness.

The doctor-patient relationship, or "clinicians without a pattern"

Within disciplines such as MG, made up, as Topol would say, of "*clinicians without patterns*," the doctor-patient relationship, context data, the uncertainty of situations, psychological state, and patient values are fundamental: patients' hopes, priorities, previous experiences, expectations, economic and family factors, religious beliefs, the desire or weariness to exist, the need for reassurance and containment, the ability to cope[2], and many other extraclinical factors that they always affect clinical management and modulate the actual implementation of the diagnostic, therapeutic and follow-up path.

Especially in this area, but not only, the predictivity of an AI model, while valid in theoretical terms (sensitivity, specificity, positive predictive value, ROC curve), must be validated "in the field," according to the risk-benefit ratio, the probable compliance of patients with what is recommended... in other words, the needs of patients, also considering the costs in economic terms and possible alternative actions.

A rigorous assessment is required, including the actual context of the application [17]. Above all, we must resolve what P. Keane and E. Topol define "*AI chasm*," that is, the fact that even an AI system with a very high diagnostic capacity may not be of great value if it does not prove to

2 It includes the psychological processes of adaptation in the face of critical and emotionally demanding situations, aimed at achieving a subjective perception of control, essential for planning effective responses and recovering adequacy.

improve clinical outcomes, by means of validation studies in real-world contexts [18].

A study, the result of a collaboration between the United States and China [19], used an AI program to analyze not only structured but also unstructured data, such as the observations of nurses, the words of children and parents, the outcomes of prescribed care, on the electronic health records of over 500,000 Chinese children, with an average age of 2.4 years.

The ML algorithm, which used *Natural Language Processing* (see glossary), has learned to associate the enormous amount of information with pediatric pathologies by imitating the clinical reasoning of the doctor to identify pediatric pathologies. In the validation process, the diagnoses identified were compared with those proposed by a group of expert pediatricians.

The results were surprising, and the success rates were extremely high, both for common diseases (95% for sinusitis and other respiratory infections) and for the most serious ones (97% for asthma attacks, 93% for bacterial meningitis, 90% for mononucleosis). The model could potentially be useful, especially for the training of young doctors, who in a parallel validation study of the tool, obtained lower results compared to AI [20].

In any case, it is essential to establish the objectives that are significant for people and to educate citizens in critical scientific knowledge (see box).

Risk communication: a social construction

In general, citizens lack the"critical" scientific knowledge capable of making them more aware of making autonomous choices regarding their own health. The patient has difficulty distinguishing between cause, necessary and sufficient, and probability, a random and nondeterministic concept. People generally think in terms of linear and rigid causal links, modulated by daily experience, heuristic cognitive devices that allow them to reach the solution of the problem in a pragmatic way with the greatest saving of time and energy.

The risk is, therefore, usually perceived at an extremely simplified level of coding; a therapy, a treatment, or a lifestyle are considered according to an all-or-nothing logic, or dangerous or safe [21]. Such biological, cognitive mechanisms, able to reassure and give simple answers to complex problems, probably emerged during the evolutionary process as defense tools, to the detriment of the objectivity of perceptions.

continued

(*cont'd*)

The involvement of the patient and the generation of a realistic perception of the level of risk are, therefore, the fundamental objective, the indispensable premises both for the protection of patients against reactions of denial or alarmism and to improve adherence to preventive measures. In fact, while sick people generally require medical intervention, in the case of a risk condition, the intervention is generally unsolicited. It must, therefore, be shared, even more than in the presence of disease, with the patient informed about the benefits and possible disadvantages.

As stated by A. Santosuosso: "*There is no end of medicine that does not collapse miserably in the face of the last patient who does not make it his*" [22].

Furthermore, while the patient directly perceives the consequences **of the disease**, he has a much more partial, and above all, variable perception **of risk**; what is acceptable for one patient may not be so for another. The path must be adapted to the individual person in its systemic, rational, and nonrational aspects. In fact, while the doctor tends to consider the risk in a technical sense, as a probabilistic event associated with multiple variables, to transform the person into an aseptic probability profile, for the patient, the risk is, as V. Andreoli states, "*Something not happened but that could happen... a non-existent that could exist... medicine of what doesn't exist, of what one thinks could be.... the dramatic world of the possible*"... ... the possibility of the occurrence of unpredictable events, able to awaken atavistic fears [23].

There is, therefore, an incommensurability between scientific logic and the psychology of the common person, which leads to behaviors which irrational on a scientific level are, however, understandable if placed in a particular context. In fact, the risk is perceived as such in reference to a mental framework, which, built on the basis of expectations and experiences, involves aspects of an emotional-affective nature. The apparent irrationality of choices depends on subjective hierarchies, contingent emotional situations, previous experiences, prejudices and preconceptions, culture, and social and family context [24].

For example, there are no rigorous studies able to demonstrate that greater prognostic precision, such as being able to tell the patient that "*there is a 27.6% probability that your smoking habit can cause cancer or heartdisease,*" causes a change in habits from more general information, such as "*there is a strong likelihood that your smoking habit could cause cancer or heartdisease*" [25].

Epidemiology in wonderland

In the epidemiological field, data flows are particularly useful as they allow taking a photograph, often instantaneous, of a certain phenomenon at a given moment, but do not allow for grasping the aspects related to the interactions between citizens/patients and the context often difficult to represent and explicitly express in terms of digitization, due to the possible presence of confounding variables and spurious correlations. Therefore, validation by external sources is necessary, in order to not arrive at wrong causal inferences that could lead to the subtraction of resources from interventions of proven effectiveness.

As the epidemiologist R. Saracci states, the value of *big secondary data* is not in their breadth but in the validity of the path that led to their measurement, the basis for estimating the internal and external validity of any research. *"The deluge of data cannot make the scientific method obsolete."*

The same precision medicine, which aims to use big data from various sources to analyze the state of each individual for predictive, diagnostic, and therapeutic purposes, in reality, according to Saracci, proposes unattainable objectives in practice.

The only possible *decision-making* tool is currently the stratification of disease risk, in terms of prognosis and response to treatments, for groups of subjects as homogeneous as possible, considering the contexts and territorial differences.

For a long period of time, interventions on populations, subpopulations, or groups will still be indispensable, favoring, especially in the pharmacological field, the response of the "average patient" (knowing that in reality, it does not exist), sick or healthy, with respect to the particular case. In fact, the predictive capacity at the cohort level does not necessarily correspond to an equally valid model in the single individual, so there remains an important share of uncertainty. A model can be valid for screening but not for diagnostics.

It is, therefore, necessary to use wise epidemiology, able to use data flows as a tool for public health knowledge, in particular, to limit the large health *gaps* between different countries and within the same populations [26].

In conclusion...

To make the DL models more generalizable and less susceptible to bias, a standardization of the data to be collected (patients of various ethnicities, languages, and socioeconomic status) and their effective integration in common formats for the different tools and IT systems is necessary, on the model of DICOM standards (*Digital Imaging and Communication in Medicine*), a platform for digital data management [27]. The same AI can also be used to reduce the risk of bias, for example, by means of "red flags" that signal situations at risk to clinicians [1].

SWOT analysis carried out by Luciano De Biase, which summarizes the opportunities and limits of AI [28]

SWOT Analysis	
Strenghts – Riduzione potenziale di errori medici e infermieristici – Velocizzazione dei processi – Accuratezza delle diagnosi	Weakness – Necessita di data base di riferimento molto grandi – Dipendenza dei risulati dalla qualita dei dati – Lunghi tempi per la validazione dei processi
Opportunities – Nuove applicazioni in espansione – Tecnologie in crescita – Personalizzazione di diagnosi e cure	Threats – Privacy a rischio – Riduzione occupazione – Necessita di aggiornamento continuo dei programmi

References

[1] Peterson ED. Machine learning, predictive analytics, and clinical practice. Can the past inform the present? JAMA 2019;322(23):2283–4. Available from: https://doi.org/10.1001/jama.2019.17831.

[2] Santoro E. https://www.agendadigitale.eu/sanita/sanita-digitale-troppe-promesse-fallite-che-ce-da-imparare/.

[3] Food and Drug Administration. FDA permits marketing of clinical decision support software for alerting providers of a potential stroke in patients. https://www.fda.gov/news-events/press-announcements/fda-permits-marketing-clinical-decision-support-software-alerting-providers-potential-stroke.

[4] Angus DC. Randomized clinical trials of artificial intelligence. JAMA 2020;323(11):1043–5. Available from: https://doi.org/10.1001/jama.2020.1039.

[5] Wijnberge M, et al. Effect of a machine-learning-derived early warning system for intraoperative hypotension vs standard care on depth and duration of intraoperative hypotension during elective noncardiac surgery: the HYPE randomized clinical trial. JAMA 2020;323(11):1052–60. Available from: https://doi.org/10.1001/jama.2020.0592.

[6] Cabitza, et al. Potential unexpected consequences of the use of oracular artificial intelligence systems in medicine. Recent Prog Med 2017;108:397–401.

[7] Rao A, et al. Sizing the prize: what's the real value of AI for your business and how can you capitalize? PwC 2017. Available from: https://www.pwc.com/gx/en/issues/analytics/assets/pwc-ai-analysis-sizing-the-prize-report.pdf.

[8] Gianfrancesco MA, et al. Potential biases in machine learning algorithms using electronic health record data. JAMA Intern Med 2018;178(11):1544–7.

[9] Maddox TM, et al. Questions for artificial intelligence in health care. JAMA 2019;321:31–2.

[10] Parikh RB, et al. Addressing bias in artificial intelligence in health care. JAMA 2019;322(24):2377–8. Available from: https://doi.org/10.1001/jama.2019.18058.

[11] Yasaka K, Abe O. Deep learning and artificial intelligence in radiology: current applications and future directions. PLoS Med 2018;15(11):e1002707. Available from: https://doi.org/10.1371/journal.pmed.1002707.
[12] Esteva A, et al. Dermatologist-level classification of skin cancer with deep neural networks. Nature 2017;542:115–18.
[13] Challen R, et al. BMJ Qual Saf 2019;28:231. Available from: https://doi.org/10.1136/bmjqs-2018-008370 23.
[14] Wang F, et al. Deep learning in medicine - promise, progress, and challenges. JAMA 2019;179:293–4.
[15] Caruana R. et al. Intelligible models for healthcare: predicting pneumonia risk and hospital 30-day readmission. Proceeding of the 21th ACM SIGKDD international conference on knowledge discovery and data minings; 2015, p. 1721-30.
[16] https://it.wikipedia.org/wiki/Fallacia_del_cecchino_texano.
[17] Shah NH. et al. Making machine learning models clinically useful. JAMA 2019; august 8; E1-E2.
[18] Keane P, Topol E. With an eye to AI and autonomous diagnosis. NPJ Digt. Med 2018;1:40.
[19] Liang H, Tsui BY, Ni H, et al. Evaluation and accurate diagnoses of pediatric diseases using artificial intelligence. Nat Med 2019;25:433–8. Available from: https://doi.org/10.1038/s41591-018-0335-9.
[20] Santoro E. https://www.agendadigitale.eu/cultura-digitale/lalgoritmo-e-un-buon-pediatra-cosi-migliorano-le-diagnosi-dellai/.
[21] Redelmeier DA, et al. Understanding patients'decisions: cognitive and emotional perspectives. JAMA 1993;270:72.
[22] https://www.slideshare.net/csermeg/informazione-e-consenso-massimo-tombesi.
[23] Andreoli V. My crazy. Memories and stories of a doctor of the mind. Rizzoli, Milan.
[24] Risk. communication: a social construction. Recent Prog Med 2020;111:1–5.
[25] Emanuel EJ. Artificial intelligence in health care. Will the value match the hype? JAMA 2019;321:2281–2.
[26] Saracci R. Epidemiology in wonderland: big data and precision medicine. Eur J Epidemiol 2018;33:245–57.
[27] He J, et al. The practical implementation of artificial intelligence technologies in medicine. Nat Med 2019;25:30–6.
[28] De Biase L. Artificial intelligence in medicine. Digital Health: from doing to treating. http://www.cdti.org.

CHAPTER

4

The cognitive biases of the digital world that are transforming our minds

The term bias literally means "oblique line"; in psychology, it is effectively used to indicate erroneous thoughts or judgments that provide a partial, flawed, or otherwise inadequate view of reality [1].

The digital world, and in particular, artificial intelligence, are highly attractive to our minds, but without our being aware of it, they alter the way we face problems and solve them, generating bias that we are not aware of.

Let us retrace together some mental processes created by digitization that have profoundly changed our way of thinking and feeling.

The enlarged mind

The foremost important feature of the digital system, from the Internet to artificial intelligence, is that it constitutes a great resource, a true enhancement of the human mind; various scholars have coined the term "enlarged mind"; this enlargement, however, can have very positive effects if the human mind uses digital rationally and critically, but it can "atrophy" the human mind if we lazily resort to digital for every trivial problem; the phenomenon has already been documented for simple mathematical calculations [2].

From sequential man to simultaneous man

In the traditional reading of texts, man generally reads, reflects, at times with a healthy critical spirit, and memorizes (**sequential man**).

DOI: https://doi.org/10.1016/B978-0-443-14054-9.00021-1

With the Internet, the fast approach is becoming increasingly widespread; since Internet search engines work much faster than us, we accept what we see on a screen as true. Even the most demanding people rarely consult more than one site when searching for an answer; thus, a new anthropological subject was created; **the simultaneous man** [3], who, if he has a problem, consults the Internet and chooses the answer most convenient or most consistent with his own beliefs (this is called the "phenomenon of the *Minimal Group Paradigm*") [4].

The illusion of digital omnipotence

The most widespread illusion is that digital is omnipotent, omniscient, and does not make mistakes; in reality, the system is powerful and precise within the boundaries that programmers have outlined and the intrinsic limits of "neural networks," some of which are unknown to the inventors themselves and manifest themselves during use. If programs and networks have limits and errors, the system does not have critical functions; it sometimes continues to err to extreme consequences.

The overconfidence

An important bias related to the previous one is the well-studied "overconfidence." In medicine, it is the most frequent mode of error for experienced doctors [5]. Overconfidence is the overestimation of one's abilities, which reduces precautions and leads to hasty actions with an increased probability of error. In the IT world, overconfidence may affect IT specialists who believe they have produced perfect systems, but more often, it affects users, who experience digital information as an unlimited extension of their mind and delegate critical, self-critical and corrective functions to digital data that rarely can perform.

Meta-ignorance

Another serious bias is meta-ignorance, known in cognitive psychology as the "Dunning–Kruger effect," named after the first researcher who studied it in depth [6].

It is a very widespread phenomenon, easily identifiable in the history of human thought; the more superficial our knowledge is, the more we presume to know the problem well; we often overestimate small discoveries or trivial intuitions.

Even this bias can affect both computer scientists and digital users, but the former are increasingly aware of their limitations, while users overestimate digital media and systematically underestimate their ignorance.

The law of least effort

A final serious limitation closing the circle is the law of least effort [7], according to which we try to obtain the best result with the least possible effort and we accept the easy answer when this is simple and fast and consistent with our beliefs.

Conclusion

When we approach the digital world, we are regularly bewitched by it and we experience it as an omnipotent means to strengthen our abilities; however, the systematic and uncritical use of digital information gives us an illusion of strength and intelligence; on the contrary it can make us less and less intelligent and above all less and less capable of critical and creative thinking.

References

[1] Galimberti U. New dictionary of psychology. Milan: Feltrinelli Ed.; 2018.
[2] Clark A, Chalmers DJ. The extended mind philosophy of mind: classical and contemporary readings. Oxford University Press; 2002.
[3] Simone R. The third phase. Forms of knowing that we are losing. Bari: Laterza Edit.; 2002.
[4] Tajfel H, Turner J. Social identity theory. Bristol University; 1980.
[5] Croskerry P. Achieving quality in clinical decision making: cognitive strategies and detection of bias. Acad Emerg Med 2002;9(11). Available from: http://www.aemj.org November.
[6] Dunning D. Advances in experimental social psychology 2011;44:259–62.
[7] Kahneman D. Slow and fast thoughts. Milan: Mondadori Edit.; 2013. p. 77–80.

CHAPTER

5

Chance/case or chaos: why is it important for us to know and distinguish them

Throwing a simple die offers us a wonderful example of the occurrence of two very different phenomena but very important and frequent in natural events and our life: Chance and Chaos.

CASE/CHANCE: when we roll a dice, a number of dots between 1 and 6 appear on the upper face; the probability of appearing for each number is one out of six; if the dice is not loaded, the succession of numbers will be completely random, and therefore, susceptible to probabilistic estimates; the more correct the higher the number of throws; with higher number of throws each face of the dice will appear about once out of six times.

From CASE/CHANCE to CHAOS: the throwing event is a process influenced by several factors; the material from which the dice is made, the force impressed in the throw, the angle of the wrist, forearm, and shoulder, the sliding plane of the dice and others. The combination of all these different variables and their close interdependence entail surprising consequences; minimal variations in one of the elements involved are enough to obtain completely different results.

The falling and rolling dice is a simple but interesting example of a small chaotic system [1].

The Mathematics of Chance and Chaos: mathematicians have been studying chance for centuries and have developed increasingly precise and reliable tools, such as the calculation of probabilities, statistical sampling, etc. As far as chaotic events are concerned, only with Poincarè, the mathematician, at the beginning of the 20th century was it fortuitously realized that it was possible to study and understand chaos (in part!). Also, the mathematics of entropy was thus developed, which allowed to predict with a certain approximation chaotic events such as meteorological ones [2].

DOI: https://doi.org/10.1016/B978-0-443-14054-9.00010-7

Chaos and case/chance: concluding reflections

Chance and chaos are very important mathematical-physical concepts as they allow us to understand natural and also psychosocial manifestations that are otherwise incomprehensible. The boundaries between the two phenomena are not always clear, but it is very important to keep them distinct as they have very different dynamics and outcomes. Let us recall, as an example of chaotic systems, the extreme weather events, once catastrophically unpredictable, today, thanks to the mathematics of chance and that of chaos, are partly predictable, and therefore, limiting in their devastating consequences.

We dwell on these abstract and perhaps abstruse concepts because, despite the thrill of neural networks and all artificial intelligence, physicists and mathematicians continue to be cautious; even the best computer devices are not always able to distinguish between chance and chaos, and so they are wrong.

Let us not trust the machines too much and do not underestimate our little big minds! [3].

References

[1] M. Li Calzi, The mathematics of uncertainty Il Mulino edit, Bologna, 2016.
[2] M. Malvaldi, S. Marmi, Chaos Il Mulino edit, Bologna, 2019.
[3] D. Ruelle, Case and Chaos Bollati Boringhieri edit, Turin, 2013.

SECTION 2

The world of sensors

CHAPTER

6

A tsunami of data: when the data is perhaps too much

Any physical, chemical, or biological entity and also a large part of psychic and socioeconomic processes can be described by means of an appropriate combination of data.

Medicine, even in its biopsychosocial dimension, does not escape this "law of data"; all this means that these are increasingly important but also more and more numerous prolems.

Semeiology, however, teaches us that an excess of data, if not properly ordered and interpreted, generates "noise" that takes us away from the truth and drags us into the swamp of uncertainty.

In this chapter, we try to provide some basic knowledge of data management in medicine.

One of the main causes of the renewed potential of AI is the exponential growth in the amount of digital data available, the so-called "big data," for which in reality, there is no clear and certain definition. In general, we mean a set of data that is so large and complex that it is difficult to process them with management tools or traditional applications. Moreover, as stated by Luciano Floridi, one of the most authoritative voices on the philosophy and ethics of information, conceiving big data starting from limited tools that suggest a problem of circularity, the data are too large in relation to our current computational capabilities. Beyond the philosophical aspects, the real problem is having information with added value, capable of being used for the advancement of knowledge, the improvement of health, and for some, the creation of wealth; a problem, therefore, above all intellectual and noncomputational power [1].

The large amount of data makes human cognitive processing capacity even more difficult. For this, heuristics are used, which in some cases cause evaluation or judgment bias. For example, in the case of the

DOI: https://doi.org/10.1016/B978-0-443-14054-9.00005-3

infodemic, when the "contagion" from the excess of misleading formations can affect the effectiveness of the public prevention measures adopted by the authorities [2].

How to Classify Data

The 3V + 1 model

An attempt to define big data is the concise description of their characteristics in the so-called "3V model" (volume, variability, velocity), which later became 4, with the addition of **Truthfulness**, correlated with the actual reliability of the same.

Moreover, the model does not completely clarify the term and leaves room for ambiguity.

Volume: reference to the enormous quantity of information, which is inevitably subject to a subjective evaluation and the relative absence of a quantitative threshold

Variability: heterogeneity of quality and news offered by the collected data: structured, semi-structured, unstructured, text, media, audio, video, sensors, Gps

Velocity: speed of collection, processing and sharing, but also often obsolete utilities

Big data is the future of medicine but also of its past. As early as 1964, inventor Zworykin had signaled that the amount of medical information that was accumulating was exceeding the cognitive abilities of health professionals. It was believed that electronic memories would prove useful supplementary tools for human memory. More specifically, Zworykin had predicted that all medical knowledge could be contained in a 10^{13}-bit memory unit [3]. A data that now makes us smile given that a single PET occupies up to 108 and every day a massive amount of digital information is produced, estimated at around [3] exabytes (10^{18} bytes, i.e., 3 followed by 18 zeros) only for devices connected to the internet. Remember that a single byte (set of 8 bits, whose value can be 0 or 1) allows the description of values between 0 and 225 (i.e. 2^8) with which to encode alphanumeric characters and that 200 characters are considered a significant quantity, for example, in linguistic terms. The sentence you just read matches exactly 294 characters, spaces excluded. To give an idea of the order of magnitude, our brain is estimated to have the memory capacity of a hard drive equal to 2.5 petabits [4].

In 1–2 years, a greater volume of data would be produced than those accumulated in the course of human history, with a total volume of 4.4 zettabytes in 2014 and 44 zettabytes according to projection estimates referable to the early years of 2020 [1]. We are in the era of the zettabytes.

It is often said that data is the new oil (*data as oil*), but in reality, this is not true. In fact, while oil is a limited resource, data is renewable indefinitely, destined to grow rapidly and continuously since the same management devices are among the main sources in a self-feeding cycle. Their value would therefore be 6 to 10 times greater than that of financial data [5].

The data sources, the object of treatment through algorithms and advanced techniques with high computational power to identify correlations, trends, and models (the so-called "data mining"), are many (often not communicating); high resolution, static, and dynamic medical images (with an enormous increase in the amount of information to be processed), personal medical devices, genomic sequencing, electronic medical records, administrative data, electronic registers, social networks, literature data, books, and libraries.

The so-called "datanami," the tsunami of data, the information tide, due to its size and complexity, has and will have a strong impact on the medical profession in areas of fundamental importance such as predictivity, diagnostics, and the doctor–patient relationship. The data are necessary not only for the initial training of the algorithms but also, subsequently, for the further improvement/validation of their performance.

Big data, despite the presence of numerous open questions, has the potential for revolutionary approaches, such as real-time monitoring of the patient in the different phases of care, the description of the clinical-care history of populations, the mapping of the human genome, the prediction of pandemics and the possible impact of containment measures, the geolocation of a specific pathology [6], the evaluation of alarm signals in the context of pharmacovigilance, the epidemiology of the target subpopulations of active ingredients, the assessment of the spread of negative behaviors, such as smoking, alcohol abuse, [7] analysis of the negative effects of environmental pollution, for example, the causal association between heart and respiratory diseases with exposure to fine particles [8].

There are, however, numerous problems. As also described in the chapter "the limits of algorithms," a large volume of data does not automatically correspond to a better quality of inferences and of the applications that derive from them. Knowledge in medicine changes extremely fast, and the amount of data increases exponentially. This creates the problem of choosing the information that is truly useful and reliable to be used as the basis for processes and related decisions.

One possible solution is the use of large databases, so-called "biobanks," and international medical image registries, such as the Cardiac Atlas Project [9], the Visual Concept Extraction Challenge in Radiology

(VISCERAL) Project [10], the UK Biobank [11] and the Kaggle Data Science Bowl [12]. Experts also consider the open availability of data and their sharing as wide as possible as a priority at an institutional level while taking into account that some technologies may have specific contextual characteristics, so much so that they cannot be generalized to all populations [13].

Financial aspects are not the main reason for advancing changes in the world of health, but, historically, they have always played a key role. The fact that the real added value of AI on clinical outcomes is still unclear and that its use in practice is still in its infancy is slowing funding in many countries, including ours.

References

[1] Floridi L. The fourth revolution. How the infosphere is transforming the world. Milan: Raffaello Cortina Editore.
[2] Conte M. Complexity Toolbox: four tools to act. In: De Simone M, editor. The complexity of an epidemic. A contribution to several voices. Complexity Institute; 2020.
[3] Greene J, Lea AS. Digital futures past – the long arc of big data in medicine. N Engl J Med 2019;381:480485.
[4] Gallina P. The liquid mind. How machines affect, modify or enhance the brain. Bari: Dedalo Editions; 2019.
[5] Topol E. Deep medicine: how artificial intelligence can make healthcare human again. New York: Basic Book; 2019.
[6] Rosa A. A semantic approach. Chall Opport Epidemiol 20 Forw 2016;04:6–7.
[7] Paolotti D, Rizzo C. Fingerprints at the service of epidemiology. Forward 2016;4:8–9.
[8] Dominici F. Look for small numbers in large numbers. Administrative databases at the service of epidemiology. Forward 2016;4:10–11.
[9] Fonseca CG, et al. The cardiac atlas project-an imaging database for computational modeling and statistical atlases of the heart. Bioinforms Oxf Engl 2011;27:2288–95.
[10] Jimenez-Del-Toro O, et al. Cloud-based evaluation of anatomical structure segmentation and Landmark detection algorithms. VISCERAL anatomy benchmarks. IEEE Trans Med Imaging 2016;35:2459–75.
[11] http://www.ukbiobank.ac.uk/.
[12] https://www.kaggle.com/datasets.
[13] He J, et al. The practical implementation of artificial intelligence technologies in medicine. Nat Med 2019;25:30–6.

CHAPTER

7

Wearable devices and the Internet of Things

Summary

The concept of health has undergone several changes in recent decades. The classic one, proposed over 70 years ago by the World Health Organization, of *a state of complete physical, mental, and social well-being and not simply the absence of disease or infirmity* has never been a real goal and a common point of reference, being considered utopian by most observers.

If we reflect on the evolution of health systems, even in the rich countries of the West, we can only see that the prevailing concept of health in advanced countries increasingly refers to exclusively biological parameters. The enormous development of sensors and remote control devices undoubtedly allows a huge leap in quality in some sectors, such as cardiology and diabetology, but it helps to spread a conception of normality not as the qualitative data "feel good" but as a sum, or calculation, of parameters set by the experts.

In this chapter, we provide an updated overview of these tools, inviting readers to always make a critical and comparative evaluation of the data they provide.

In the *Hundred Person Wellness Project* [1], 100 healthy volunteers were followed intensively for 9 months by continuous monitoring of sleep, physical activity, and heart rate, associated with a group of about 100 biochemical tests on blood, saliva (genome), urine, feces (microbiome) every 3 months. The study, with no control group, found something abnormal in almost all participants, from reduced vitamin D

continued

DOI: https://doi.org/10.1016/B978-0-443-14054-9.00024-7

(*cont'd*)

levels to prediabetes. This was followed by the proposal of a paid platform to carry out the same exams under the supervision of a *coach* and the promotion of a large-scale study on 100,000 healthy people [2].

Digital culture is causing radical transformations in our society. According to many experts, in the future, we will no longer talk about "personalized or smart medicine" but only about medicine [3]. Technologies are transforming the times and spaces of health, in particular through the continuous monitoring of people, at home, during leisure time, and also in the workplace, recording the events that occur during the so-called "other 362 days a year," when patients are not seen by health professionals [4]. The smart home device market is evolving exponentially, going from $ 6.8 billion in 2016 to $ 14.7 billion in 2017, with forecasts exceeding $ 101 billion by 2021 [5].

Real-world health data, many of which were previously unavailable, can be obtained directly from the producers/owners themselves, who, in about 20 years, will perhaps end up taking control of it, permanently changing the traditional doctor-patient relationship [6].

Moreover, one of the dangers is that described in the initial box, a scenario that could also arise in daily practice. New technological developments are, in fact, orienting medical culture more and more toward measuring health and well-being. The keywords are increasingly calculability, quantification, and controllability. Life itself risks becoming a matter of medicine, as it can be objectified in medical terms.

The new technologies, useful for studying pathological conditions, are currently used above all in conditions of well-being due to the importance that many people attribute, in today's culture, to monitoring their health conditions, creating a sort of *data-driven world* [7].

Wearable devices

The main tools of digital medicine, in addition to the technology of electronic records, *online* services (consultation of diagnostic or specialist reports), and the tools used for interaction with patients (*e-mail, sms, social networks*) are wearable devices, consisting of one or more biosensors, whether or not equipped with artificial intelligence (AI), which, inserted on clothing such as watches (*smartwatch*), t-shirts, shoes, trousers, belts, bands (*smart clothing*), glasses (*smart* glasses), they can detect and measure different biological parameters [heart rate (HR), spirometric variations, oxygen saturation, body temperature, blood pressure, glucose, sweat, breath, brain waves] and provide information on lifestyle (physical activity, sleep, nutrition, calories consumed).

WDs can send initial *feedback* to the person wearing them, generally through *smartphone* applications (apps), and then to the relevant *cloud* service, where they are organized, by the use of AI algorithms, to be available and interpretable by the user or by other figures, for example the attending physician or his *team*.

WD, also known as "intelligent agents," are a category with a high potential for expansion, whose market is constantly growing, fueled by various factors, such as much lower costs and improvements in technology, which have made it possible to miniaturize the electronic components, making the devices lighter and smaller in size so that they can be worn or integrated into clothing [8], becoming in some cases real *status symbols*.

The sale of digital products directly to consumers is also increasingly widespread, especially devices for fitness, sleep monitoring, and the detection of cardiac arrhythmias. In the United States, 252 (21%) of digital health industries have a direct-to-consumer strategy, with a turnover of 7.3 billion dollars from 2011 to 2019. This rapidly expanding market can theoretically be useful for people who lack the ability to access care due to economic, social, or geographical barriers. For example, web-based applications for screening people at high risk of psychosis, telemedicine services or devices to detect arrhythmias such as atrial fibrillation (see chapter) have found particular favor with the public. However, there are numerous perplexities, such as the lack of filters as regards reliability and possible incongruous use of the devices. Extreme caution on the part of institutions, clinicians, and researchers is, therefore, essential, in particular, rigorous clinical validation studies are [9] indispensable.

Fields of application

The continuous monitoring of biomedical parameters would be able, according to the intentions of those who propose them, to educate patients, and citizens in general, to adopt healthy behaviors and to change their lifestyle, with a view of not only promotion and health prevention, but also, as far as the sick are concerned, diagnostic activities, treatment management, and rehabilitation.

Wearable technology makes it possible to carry out diagnostic tests and to monitor individual body functions in the context of care, even in remote places (*Point-of-Care Testing*), ideally being a great resource, especially for low-middle-income countries, keeping in mind that over 80% of the adult population currently has *smartphones* and broadband connectivity.

A sort of "virtual flattening" of the earth is therefore conceivable by means of equal access to these technologies, an integration between digitization and democratization [10]. However, these possibilities are still at a very early stage, even if increasingly innovative biosensors are being developed and regulated for clinical use. For example, it is already possible to perform a digital, physical examination remotely using a device that works as a thermometer and stethoscope and which can send the breath recording to the doctor's office to assess the nature of a cough [11]. This will become increasingly important due to the limitations that will characterize the activity of doctors due to the pandemic from COVID-19 underway as we are writing.

The "sensorization" is now part of the daily life of many people, especially of those who actually need it less: young people, on average well-off, technologically competent, and already strongly oriented to use technology. The objectives of "healthy" users are different: from the simple recording of data by subjects who already adopt a healthy lifestyle and simply want to quantify their progress (number of daily steps, maximum, average and instantaneous walking speed, time weekly dedicated to moderate physical activity, *cardiorespiratory fitness*), for use aimed at improving one's health, psychological and emotional well-being, sociability and interpersonal skills, productivity, and professional performance. The concept of life as a phenomenon that can be measured using specific tools has even determined the development of a global network of enthusiasts within a broad cultural movement called "quantified self," whose slogan is: "self-knowledge through numbers" [12].

The most classic areas

"Practical" applications of the use of new technologies also exist in a more strictly clinical setting, for example, in the management of cardiovascular diseases, from hypertension to heart failure to ischemic heart disease.

Wearable sensors, based on photoplethysmography and radar technology, can, for example, continuously and noninvasively measure blood pressure, evaluating the variations according to the patient's activities, day by day, minute by minute, in addition to the classic night-time circadian trajectories. The blood pressure data can be aggregated and displayed instantly on a *smartphone* to obtain *direct* feedback for the user.

Continuous digital technology could lead to a much deeper understanding of the blood pressure construct itself and allow a redefinition of the classic, imprecise diagnosis by describing distinct and individualized clinical phenotypes not detectable with common tools. This could allow for a truly person-centered, pragmatic, epistemological, and far-reaching clinical therapeutic approach [13]. However, there are still many uncertainties relating, for example, to the accuracy of the obtainable data, to the treatment criteria, and to the actual value in terms of improving clinical results compared to traditional therapy.

Diagnostic tests can be carried out by means of smartphones equipped with AI for various medical conditions, such as skin lesions, ear infections, and retinopathies.

Some apps are able to monitor patients' adherence to therapy, for example, AiCure, through which the patient takes a video selfie when swallowing the prescribed tablet.

A study involved a cohort of 800 subjects, overweight in 54% of cases and obese in 22%, nondiabetic, who, subjected to continuous monitoring of blood glucose for a week, showed great variability in response to the same diet. A computational algorithm integrated traditional data (food diary, anthropometric data, physiological parameters, physical activity levels) with the profile of the intestinal microbiome, creating a predictive model of glycemic response, subsequently tested on a cohort of 100 people and finally on 26 participants, randomized to different nutritional interventions. Personalized diets have been shown to be more effective in reducing glycemic fluctuations and postprandial glycemia than traditional [14] diets. Studies of this kind can be considered curious "prototypes" but do not currently allow any application to daily practice.

Regarding the management of diabetes, the development of wearable devices is still under development, although the prospects are fascinating, especially with regard to enzymatic noninvasive methods, such as contact lenses for tears and skin *patches* for sweat [15].

Currently, algorithms can be used on the basis of glycemic variations and allow better glycemic control by the patient himself. Fortunately, tools are provided in this context to predict and avoid abnormal behavior of intelligent machines, for example, the risk of severe hypoglycemia [16].

In the near future, the use of voice assistants is foreseen, which through the automatic learning of *machine learning* (ML), analyze the *trends* of the single individual, together with other covariates such as sleep, physical activity, weight, nutrition, comorbidities, therapies in place, and activate graduated, personalized therapeutic approaches aimed at the patient or caregiver.

Consensus documents have highlighted the potential of WDs in recognizing and monitoring atrial fibrillation (AF), although the accuracy of the signal, which is good in detecting arrhythmia, is not always adequate in monitoring HR. A recent meta-analysis conducted on 10 studies involving 3852 patients and 4 apps, showed that smartphone apps that use photoplethysmography signals have a high sensitivity (94.2%) and specificity (95.8%) for the diagnosis of AF. In particular, they seem adequate to exclude it from suspicious patients. However, the use of these devices as screening tools in asymptomatic populations can increase the number of false positives [17]. (see also paragraph apple watch).

The Internet of Things

The integration between *smartphones* equipped with AI, biosensors, and the so-called "home utilities" is also transforming the "where" care will be provided in the years to come, especially for patients with chronic diseases, who can hardly be transported from their homes to the hospital.

The so-called "Internet of Things" (IoT) is the network of technological devices connected to the Internet, the process by which objects, devices, and systems become "smart," i.e., equipped with software that allows them to identify themselves electronically, connect and communicate their performances, exchange data and perform processing. In particular, the IoT aims to create a *smart medical home,* characterized, for example, by the use of sensors, at floor level or wearable cameras and infrared devices, which assess the risk of falling, and in general, the validity of walking or recognize the fall itself, with warning signals for *caregivers* or reference centers connected remotely, through different mobile *gateways* (smartphone or other dedicated device) and from there to the network. With the advent of 5G it will be possible to connect devices directly to the network without any human intervention.

In the *smart home* PA, HR, body weight, urine, and feces can be analyzed by the sensors when the patient sits on the toilet. There are also devices, placed between the mattress and the sheets, capable of automatically recording sleep data and sending them via a *blue tooth* connection to an app on a *smartphone* or *tablet*. The sensor recognizes when the patient lies down and automatically starts monitoring, collecting, and analyzing sleep duration and effectiveness, HR, breathing, movement, snoring, room temperature, and humidity.

ML algorithms would be able to identify and analyze the introduction of food and combine this information with the passively monitored weight change at the floor or *toilet* level to create individualized diet plans, also taking into account calorie consumption. The accuracy of food introduction is, however, difficult to ensure. For example, for the common user, it can be difficult to identify and quantify the individual components of complex foods [18]. There are also apps, which by means of image recognition, establish the caloric and nutritional content of foods [19]. Facial recognition sensors, videos, and video mirrors capable of analyzing the swallowing of the drug itself are under development.

Instead, devices are already available that can track the air breathed with inhalers for patients with asthma and COPD [20].

In one study, the analysis, by means of ML, of the monitoring data, through wearable devices, of saturation, HR, number of daily steps, body temperature, and respiratory rate, showed, in patients with COPD, to accurately predict (AUC 0.87) severe exacerbations with a mean advance of 3 days. The most accurate predictors were HR and the number of steps [21].

Other telemedicine systems would also be able to predict the exacerbation risk of COPD patients. The most significant parameters were FEV1, oxygen saturation, and body weight [22].

Applications for the calculation of the risk of exacerbation of obstructive pulmonary diseases on the basis of the values of environmental pollution and concentration of pollen in the air are also being advanced.

In the future, it is possible to imagine a sort of virtual coach, associated or not with a patient avatar, able to create a holistic and personalized approach through the integration of all the patients' multimodal inputs, deriving from behavioral analysis, genomics, biosensors, intestinal microbiome, environment, physiological activities, drugs, blood-chemical tests, family history, cognitive status, with continuously updated medical literature.

According to the writer and philosopher Eric Sadin, the risk is that a sort of "algorithmic assistantship" provokes the disintermediation/dethronement of the doctor, making the obsolete consultancy phase useless to offer constant monitoring, offered for a fee by the medical industry. There is a risk of a gradual ban on the skills of doctors and their cognitive faculties for the benefit of devices called upon to perform not a complementary but exclusive function, validating the confiscation of medicine by the digital industry in association with the world of drugs [23].

The clinic of the future

Sadin describes the start-up Forward whose ambition is to create the clinic of the future, a space intended to collect the greatest number of information on each client who, during registration, is subjected to a series of tests: weight control, height, temperature, heart rate and blood pressure, Computerized tomography (CT) scan, blood, and saliva tests for genomic analysis. An AI system should monitor the conversations between patient and doctor to detect all the elements useful to outline the profile. Forward provides the devices connected to the patient: bracelets, scales, tools for sleep analysis, and performing instrumental tests such as the electrocardiogram. The collected data are studied remotely by an algorithm, which in the event of an anomaly, can schedule a visit

continued

(cont'd)

or proceed to analysis and tests. The company, upon subscription, also offers the possibility to take advantage of a theoretically unlimited number of views and vaccinations, to be able to contact nutritionists and to receive a supply of drugs. The founder hopes to open these facilities in the United States and abroad and says he "wants to rebuild everything from scratch" [24].

Wearable ... but ... how much "able"? that is, very comfortable ... but ... do they work? ***What clinical trials tell us.***

Most of the clinical studies have looked at the impact of wearable technology on particular tasks for the achievement of a specific goal, typically regarding lifestyle, typically physical activity, and weight loss. For example, a meta-analysis examined 27 trials but without detecting significant changes, through the use of remote monitoring, of none of six predetermined *outcomes*: body mass index (BMI), waist circumference, percentage of adipose tissue, systolic and diastolic blood pressure. The studies were highly heterogeneous in terms of protocol, *device* types, and outcomes considered. The most effective interventions, as in other studies [25], were based on the personalization and supervision of qualified personnel. The authors report the need for more evidence of efficacy before implementing the remote technology in the clinical [26] setting.

A trial conducted on elderly patients found that the use of pedometers, combined with nursing counseling, allowed for increased physical activity [27]. A *review* identified 35 studies, mostly randomized clinical trials, in 77% of which improvement in activity performance was confirmed [28]. A 2017 meta-analysis of 21 trials showed limited effects [29].

The *MyHeart Counts mobile CV health* study, carried out by Stanford University, demonstrated the feasibility of smartphone-based research in the cardiovascular field and the possibility of analyzing, quickly, on a large scale and in detail, some parameters such as activity, sleep, and physical fitness, diet, state of well-being, perception of risk, work and leisure activity, sleep, cardiovascular parameters. In particular, the study confirmed the poor correlation between measured and self-reported physical activity. The study, however, has several serious limitations recognized by the authors themselves. The first is the over-representation of classic smartphone users, *young* people (average age 36 years), and males (82%). Furthermore, as in other studies of this kind, there was [30] reduced participation over time, with a loss of data [1]. In fact, approximately 49,000 subjects agreed to participate, but only 20,345 individuals (41.5%) completed 4 of the 7 days of registration and 4552 subjects (9.3%) every 7 days.

In the IDEA (*Innovative Approaches to Diet, Exercise and Activity*) randomized controlled trial conducted on 470 young adults, with BMI between 25 and 39, over a period of 2 years, the addition of wearable technology to behavioral interventions, compared with treatment standard, it was associated with less weight loss [31]. The lack of consistency of users over time was also limited in this study: 32% of subjects stopped using the devices after about 6 months, and 50% after a year. The reasons for discontinuing use are essentially forgetting to wear the device, discomfort during monitoring, lack of esthetic value, loss of interest in the method, and satisfaction for having achieved the desired results [32,30].

It is, therefore, essential that wearable technology is included in an adequate and sustained strategy of involvement/motivation of patients to maintain healthy behaviors or modify negative ones.

According to some experts, the inconveniences attributable to constant monitoring could be limited by reducing the surveillance periods and limiting them to only one pathological condition at a time [33]. SM Patel proposes other engagement *strategies* [34]:

- reimbursement of the cost of the instruments once the results have been demonstrated in clinical terms, as well as, due to their use, savings on the costs of healthcare services
- individualization of specific objectives such as ideal weight or the daily amount of kilometers traveled
- greater simplicity in their use for less "technological" subjects
- greater use of feedback *tools* capable of stimulating the user.

For this purpose, industries and researchers have created platforms to enable performance improvements and strengthened people's commitment through sharing among peers, including through gamification techniques, with the publication of results *online*, competitions between participants, in some cases organized in groups, with economic incentives for the results achieved [35,36,37]; however, with conflicting results, in some cases negative, especially as regards the initial phase, in the elderly and in low-income subjects [38].

For an evaluation of the clinical applicability of some WDs for the promotion of physical activity, see a recent focus of the AHA, in which, moreover, multiple difficulties are described, especially for the heterogeneity in activity measurements and validation studies, as well as for the limited integration with the data available by health professionals, in particular with electronic medical records [39]. Comparisons between different physical activity monitoring devices also showed wide variations in accuracy between different devices, with a margin of error of up to 25% [40,41]. Accuracy is particularly reduced during intense physical activity due to motion artifacts [42].

A *review* analyzed nine studies concerning the detection of the risk of falling by WD, usually positioned on the belt. Moreover, the results, which demonstrated good sensitivity (≥85.7%) and specificity (≥90%), were mostly related to artificial scenarios, with laboratory simulated events on young adults. Extensive validation studies on real populations are, therefore, indispensable, capable of detecting the risk of falling, any fall, and even daily activities [43].

One wonders if efforts to address this issue would not be better directed towards greater education of patients and *caregivers in* standard measures to prevent falls, which have proven effective in many studies.

Scientific works have focused mainly on the feasibility of *sensing*, no large-number trials have shown a positive impact on clinical outcomes25. The technological possibilities are still under development, there are many doubts regarding reliability and validity, security and privacy. The possible "recording artifacts," the lack of standardization and calibration of the devices, the nonoptimal estimate, for example, of physical activity and energy consumption, are still to be resolved [44].

Studies on the use of WD, in general, involve small numbers, are short-lived, often conducted by the developers themselves or by structures with direct financial interests and not by independent researchers, are not rarely based on the results of subjects describing their experiences, not conducted in real-world settings. No data on possible adverse effects are available. Studies against placebo are lacking, so much so that it is believed that part of the positive responses depends on a "digital placebo effect," also given the close relationship between people and their *smartphones*. There are no definitive analyzes on the cost/effectiveness ratio in practice, on the actual ability/willingness of people to take charge of their health directly.

There are difficulties in acquiring data by elderly people, or in any case, inconveniences, such as not to make intensive monitoring feasible (see box).

Some studies have used a complex integration of telemedicine tools but have not led to significant results. The *Whole System Demonstrator* trial provided practically zero results in terms of cost/effectiveness [45].

A review of the *Health Innovation Network*, commissioned by the British Health Service and limited by a series of methodological problems, found no evidence of efficacy in the management of common chronic diseases through video consultations, while remote monitoring provided good results in the control of diabetes, hypertension, heart disease, and COPD. Texting has shown beneficial effects in monitoring diabetes, short-term smoking cessation, and adherence to complex therapies, such as antiretrovirals. The apps have had inconclusive results [46].

Wearable technology—possible advantages and benefits

- Central role of the patient
- Patient-friendly *use* (for "digital natives")
- Facilitation of home care
- Limitation of hospitalizations
- Collection of previously inaccessible data
- Reduction of the risk of error in data collection, as they are automated, with a desirable increase in efficiency and quality
- MORE CARE CUSTOMIZATION
- Improvement of lifestyles
- Improvement of clinical outcomes.

Wearable technology—main limitations and obstacles

- Lack of awareness and in general low confidence in digital solutions by the majority of patients/citizens and health professionals/institutions
- Reduced availability of device reliability studies (versus *gold standard methodology)*
- Lack of standardization and common calibration between different devices
- Lack of interoperability between different areas and health services, for example of integration between the data obtained and electronic medical records
- Lack of high methodological studies able to provide evidence of efficacy on clinical outcomes
- Reduced investment and poor economic incentives to support initial costs by patients and health services
- Effective usability of the data, both by the patient and by healthcare professionals (difficulty in acquiring data by the elderly due to the *digital divide*; need for specific training for professionals,

continued

(*cont'd*)

modification of the workflow to use the data provided by technologies in daily activities, generally defined in different ways)
- Reduced availability of validated data on the incremental value in terms of cost/effectiveness compared to standard assistance
- Issues related to security and *privacy*
- Negative aspects of intensive monitoring
- Possible false sense of confidence in users' self-diagnosis
- Concern about the real possibility of access to technology
- Possible further medicalization of health and life.

Reflections

Digital culture offers new potentials capable of changing the world of medicine at a specialist and territorial level. For a practice of digital culture, rigorous clinical validations, efficacy/effectiveness studies, and new types of pragmatic trials that use the principles of ML and digital interfaces to interact with electronic records and with standard methods of diagnosis, monitoring, and processing, in order to obtain a real understanding of the data, all in a strictly contextualized and personalized manner [47]. The interaction between *device* and patient is, in fact, complex; studies are needed to evaluate the types of subjects who can actually benefit from it.

Moreover, the use of WD is not only a problem of effectiveness/efficiency but of a change of cultural paradigm. The risk is that a sort of new sensory apparatus is being created, a pervasive instrumentation capable of accessing physical, social and environmental realities in ways, scales, and forms that are unprecedented in the history of humanity. In the words of Eric Sadin, "the body becomes, we become, the center of attention of the systems."

As a result of these new perspectives of investigation, we are beginning to record with new eyes and redefine the very concept of body identity. As stated by the philosopher C. Accoto, "*it is not just about tools to calculate the number of steps or the amount of calories burned but about tools with which we are building our new (idea of) human subjectivity*" [48].

New technologies generate unprecedented horizons of the possible which is apparently only theoretical and can become real in a short time. The general direction is marked, the devices will sooner or later become reliable, probably in forms that are not even conceivable at the moment, and will certainly find indications of use, at least in selected patients and contexts. The industry is investing heavily in new technologies and we know that, in the clash between interests and values, the former is destined to prevail.

However, the future is not completely determined. It is possible to intervene on its development, to try to reintroduce an explicit dialectic into the culture of medicine, a comparison of thoughts, methods, objectives, necessary to enter into the merits of the paths, the management of projects, the defense of citizens/patients, avoiding a sort of *power pointlogic*, characterized by "given" communications, *fashion* presentations of realities considered in an optimistic and definitive way.

The doctor must prepare to manage the relationship with patients who will increasingly submit the data obtained with the WDs (it is estimated that about 325,000 health-related apps are available on the main stores), with the risk of being overwhelmed by an enormous mass of information and new responsibilities, in a context of greater uncertainty and confusion, for example due to expectations placed on technology. Obviously, buying a mobile phone and wearing a *device* is not enough to change behaviors such as physical activity or nutrition. Lifestyle is a complex feature, which, for example, reflects the adaptive response to the need for integration between the culture of the individual and that of the community to which he/she belongs and is, therefore, largely influenced by the habits of society.

There is also the possibility that patients may have excessive confidence in self-monitoring and "do-it-yourself" diagnoses, which are actually not very reliable, and in any case, cannot be inferred simply from data analysis.

It is, therefore, desirable (indispensable?) A collaboration between clinicians and developers to integrate the possibilities of technology with the experience of practice to respond to the real needs of people and their real needs for care. Let us not forget that our identity is still fundamentally analog, even in an increasingly digital world. In this regard, it is interesting to note that the term digital derives from the English digit (which means digit, referring in this case to the binary code), which, in turn, derives from the Latin digitus, "finger" (in fact, with the fingers one can count the numbers). Despite the etymology, the concept of digital medicine has become an oxymoron in practical use: human touch versus its antithesis, contact versus monitoring, with an ever-increasing risk of losing the doctor-patient relationship.

Surely medicine can (will) never be virtual nor (probably) tackle only with sensors or algorithms. This certainty (or just hope?) should be the real "technological" innovation, with high added value, flexible, powerful, and economical, oriented to the real needs of people.

References

[1] Gibbs WW. Medicine gets up close and personal. Nature 2014;506:144–5.
[2] Cross R. Scientific wellness'study divides researchers. Science 2017;357:345.
[3] Lane R. Eric Topol: innovator in cardiology and digital medicine. Lancet 2016;387:1267.
[4] Rui P, et al. National ambulatory medical care survey: 2014 state and national summary tables. Available from: <https://www.cdc.gov/nchs/data/ahcd/namcs_summary/2014_namcs_web_tables.pdf> [accessed 15.05.18].
[5] Zuboff S. Surveillance capitalism. The future of humanity in the era of the new powers. Rome: LUISS; 2019.
[6] Topol E. The patient will see you now. New York: Basic Book; 2015.
[7] McKinsey Global Institute. The age of analytics: competing in a data-driven world. McKinsey Global Institute; 2016. Available from: https://www.mckinsey.com.
[8] Stefano Z. Internet of things people, organizations and society 4.0. Rome: LUISS; 2018.
[9] Cohen AB, et al. Direct-to-consumer digital health Lancet Digital Health 2020;2030057-1. Available from: https://doi.org/10.1016/S2589-7500.
[10] The Economist. Planet of the phones. Available from: <https://www.economist.com/leaders/2015/02/26/planet-of-the-phones>; 2015.
[11] Green JA. Do-it-yourself of medical devices—technology and empowerment in American Health Care. N Engl J Med 2016;374:305–8.
[12] http://quantifiedself.com/.
[13] Steinhubl SR, et al. Digital medicine. Off the cuff: rebooting blood pressure treatment. Lancet 2016;388:749.
[14] Zeevi D, et al. Personalized nutrition by prediction of glycemic responses. Cell 2015;163:1079–94.
[15] Lee H, et al. Enzyme-based glucose sensor: from invasive to wearable deice. Adv Healthc Mater 2018;7:1701150.
[16] Thomas PS, et al. Preventing undesirable behavior of intelligent machine. Science 2019;366:999–1004.
[17] O'Sullivan JW, et al. Accuracy of smartphone camera applications for detecting atrial fibrillation. A systematic review and meta-analysis. JAMA Netw Open 2020;3(4):e202064. Available from: https://doi.org/10.1001/jamanetworkopen.2020.2064.
[18] Lobstein T. Can wearable technology help patients tackle obesity? BMJ Opin 2016.
[19] Topol E. High-performance medicine: the convergence of human and artificial intelligence. Nat Med 2019;25:44–56.
[20] Kvedar JC, et al. Why real-world results are so challenging for digital health. N Engl J Med Catalyst 2017; (July).
[21] Esteban C, et al. Machine learning for COPD exacerbation prediction. Eur Respir J 2015;46. Available from: https://doi.org/10.1183/13993003.congress-2015.OA3282 OA3282.
[22] Mohktar, et al. Predicting the risk of exacerbation in patients with chronic obstructive pulmonary disease using home telehealth measurement data. Artif Intell Med 2015;63:51–9.
[23] Sadin E. Critique of artificial reason. A defense of humanity. Rome: LUISS; 2019.
[24] Marin J. Bienvenue dans le cabinet medical du future. Le Monde 2017; March 31.
[25] Martin SS, Feldman DI, Blumenthal RS, et al. mActive: a randomized clinical trial of an automated mHealth intervention for physical activity promotion. J Am Heart Assoc 2015;4:e002239.

[26] Noah B, et al. Impact of remote patient monitoring on clinical outcomes: an updated meta-analysis of randomized controlled trials. NPJ Digital Med 2018;1:20172. Available from: https://doi.org/10.1038/s41746-017-0002-42.

[27] Harris T, et al. A primary care nurse-delivered walking intervention in older adults: PACE (Pedometer Accelerometer Consultation Evaluation) – lift cluster randomized controlled trial. PLoS Med 2012;12(2):1001783. Available from: https://doi.org/10.1371/journal.pmed.1001783 PMID: 25689364.

[28] McConnell MV, et al. Mobile health advances in physical activity, fitness and atrial fibrillation. JACC 2018;71(23):2691–701.

[29] Direito A, Carraça E, Rawstorn J, Whittaker R, Maddison R. mHealth technologies to influence physical activity and sedentary behaviors: behavior change techniques, systematic review and meta-analysis of randomized controlled trials. Ann Behav Med 2017;51:226–39.

[30] Ledger D, Partners E, Scientist B, Manager P. Inside wearables. How the science of human behavior change. Endevour Partn 2014. Available from: http://endeavour-partners.net/white-papers/.

[31] Jacicic J, et al. Effect of wearable technology combined with a lifestyle intervention on long-term weight loss. The IDEA randomized clinical trial. JAMA 2016;316 (11):1161–71.

[32] Fox G, et al. Why people stick with or abandon wearable devices. N Engl J Med Catalyst 2017; (September).

[33] Muse ED, et al. Digital medicine towards a smart medical home. Lancet 2017;389:358.

[34] Patel SM, et al. Wearable devices as facilitators, not drivers, of health behavior change. JAMA 2015;313 459-50.

[35] Piwek L, et al. The rise of consumer health wearables: promises and barriers. PLoS Med 2016;13(2):e1001953. Available from: https://doi.org/10.1371/journal.pmed.1001953.

[36] Ganesan AN, Louise J, Horsfall M, et al. International mobile-health intervention on physical activity, sitting, and weight: the Stepathlon Cardiovascular Health Study. J Am Coll Cardiol 2016;67:2453–63.

[37] Patel MS, Asch DA, Rosin R, et al. Framing financial incentives to increase physical activity among overweight and obese adults: a randomized, controlled trial. Ann Intern Med 2016;164:385–94.

[38] Patel MS, Foschini L, Kurtzman GW, et al. Using wearable devices and smartphones to track physical activity: initial activation, sustained use, and step counts across sociodemographic characteristics in a national sample. Ann Intern Med 2017;313:625–6.

[39] Lobelo F, Young DR, Sallis R, et al. Routine assessment and promotion of physical activity in health care settings: a scientific statement from the American Heart Association. Circulation 2018;137:e495–522.

[40] Lee JM, Kim Y, Welk GJ. Validity of consumer-based physical activity monitors. Med Sci Sports Exer 2014;46(9):1840–8. Available from: https://doi.org/10.1249/MSS.0000000000000287. PMID:24777201.

[41] Houses MA, Burwick HA, Volpp KG, Patel M. Accuracy of smartphone applications and wearable devices for tracking physical activity data. JAMA. 2015;313(6):10–11.

[42] Spierer DK, et al. Validation of photoplethysmography as a method to detect heart rate during rst and exercise. J Med Eng Technol 2015;39:264–71.

[43] Pang I, et al. Detection of near falls using wearable devices: a systematic review. J Geriatric Phys Ther 2018. Available from: https://doi.org/10.1519/JPT.0000000000000181. PMID: 29384813.

[44] Dooley EE, Golaszewski NM, Bartholomew JB. Estimating accuracy at exercise intensities: a comparative study of self-monitoring heart rate and physical activity wearable devices. JMIR Mhealth Uhealth 2017;5:e34.

[45] Henderson C, et al. Cost effectiveness of telehealth for patients with long term conditions (Whole Systems Demonstrator telehealth questionnaire study): nested economic evaluation in a pragmatic, cluster randomized controlled trial. BMJ 2013;359:11035. Available from: https://doi.org/10.1136/bmj.f1035pmid:23520339.
[46] Health Innovation Network. NHS England: TECS evidence base review. <https://healthinnovationnetwork.com/wp-content/uploads/2017/08/NHS-TECS-Evidence-Base-Review-Findings-and-recommendations.pdf> 2017.
[47] Barrett PM, et al. Digital medicine digitising the mind. Lancet 2017;389:1877.
[48] Accoto C. The given world. Five short lessons in digital philosophy. Milan: EGEA; 2017.

CHAPTER

8

The sentry watch

Summary

"Dum Romae consulitur, Saguntum expugnatur" is a famous expression of the great Tito Livio, which literally means "While Rome is discussing, Sagunto is conquered."

We have been discussing for years with concern and passion the potential and limits of the devices that control our state of health, evaluating their potential and "adverse effects."

Now Apple confronts us with the "fait accompli" with its valuable Apple 4, of which we recognize the indisputable merits, but we also see potential important negative effects.

> ...It is technology that merges with life and becomes the digital version of J. Bentham's panopticon, the prison surveillance device that inaugurated the modern society of control and prevention at the end of the eighteenth century. And now the hundred eyes are all on the iWatch screen. Perhaps it is no coincidence that, in English, sentry guard are called watch, with the same word as watch. Thus each of us becomes a supervised overseer. ***An "iWatch me" in flesh and blood. (M.Niola) [1].***

Consensus documents highlighted the potential of wearable devices in recognizing and monitoring atrial fibrillation (AF) [2,3]. Here we intend to address the theme of the Apple Watch 4, whose advertising launch, aimed mainly at the consumer world, has made a lot of noise, not only of technological advances in general, but above all, because the latest version is able to record an electrocardiographic trace to reveal the important arrhythmia, a common cause of stroke.

The EKG App has been registered in the United States by the FDA, the regulatory body that supervises drugs and medical devices, as a medical device and also has the CE marking and authorization for the European Economic Area. In Italy, it is classified by the Ministry of Health as "Electrocardiographs" among the "Software accessory components" and identified by the trade names "EKG App" and "Irregular Rhythm

DOI: https://doi.org/10.1016/B978-0-443-14054-9.00009-0

Notification Feature" in the so-called "Class IIa" (active devices that interact with the body in a nondangerous way) [4].

The Apple Watch electrocardiogram

The Apple Watch has a photoplethysmography sensor[a], which detects the blood flow in the user's wrist. The variability of the heart rhythm is detected by means of a tacogram, which is a graph of the time between one heartbeat and another. The device can be activated easily in the Health app of the compulsorily paired iPhone. If the user wishes to obtain a trace, for example, in the case of symptoms, such as a rapid or irregular heartbeat, he must be at rest, with his arm resting on a surface or in his lap. At this point, placing your finger on the crown of the smartwatch acquires in 30 s a single-lead electrocardiogram (EKG), similar to D1, which gets stored in the Health app in pdf format and can be shared via email or message.

The activation of the app also starts a periodical passive monitoring in the background, however only in the periods in which the user does not move for a time sufficient to obtain the reading. In the event of irregular tacograms, another series is started, with a higher acquisition frequency (as often as possible, at a minimum distance of 15 min). If in a 48-h period, five out of six tacograms are classified as irregular, the wearer of the watch receives a notification of potential arrhythmia from the device algorithm, and further information in the Health app, such as the hour of the irregular heartbeat. In the case of other types of arrhythmias, for example, simple extra systoles or artifacts, the automatic diagnosis fails and the message "electrocardiogram not evaluable" appears on display.

All EKGs and other findings, including any symptoms, are searchable and can be sent to the doctor. The digital watch cannot perform a complete 12-lead EKG, and therefore, cannot diagnose other conditions, such as myocardial infarction.

The watch is also able to detect possible falls, thanks to the analysis of the trajectory of the wrist and the acceleration of the impact, and to send a notification that can be canceled or used to initiate an emergency call. In particular, if the Apple Watch detects that the user is immobile 60 s after the fall, it automatically calls for help and sends a message with its position to the emergency contacts.

The ability of the Apple Watch to detect the presence of AF was the goal of the [5] pragmatic, prospective, single-arm, open-label,

a technique based on the principle that the red color of the blood absorbs the frequencies of green light, which, by means of photodiodes, can be quantified and reported as heart rate

Apple-funded Stanford University Apple Heart Study. The study, the largest on arrhythmias ever conducted, involved over 400,000 people for 8 months, on a voluntary basis, probably curious and eager for reassurance from an innovative digital tool. The people included in the study were required to own an iPhone and an Apple Watch, and therefore, were the sponsor's customers [6]. The average age of the participants was about 40 years, with only 6% over 65. 52% under the age of 40 (22–39). Subjects with AF or atrial flutter or treated with anticoagulants were excluded.

These in summary are the results:

- During the monitoring, over 2000 people (*0.52%*) received a notification of suspected cardiac irregularity (notification rate*)*; of these, only 450 (20.8%) received a conventional EKG patch, which was monitored for up to 7 days (however, applied, on average, 13 days after reporting)
- The notification rate was dependent on age: 3.2% for those over 65, 1.3% in the 55–64 range, 0.37 in the 40–54 range, and 0.16% in the under 40 s.
- The control EKG patch confirmed AF in 34% of cases; however, only in 18.4% of those under 40.
- Positive Predictive Value of patch-confirmed notifications was *84%*.
- Of the participants who received the notification only 20.8% returned the confirmatory EKG pattern.
- The 16 adverse events reported involved anxiety problems.

The VPP is acceptable in absolute terms, not for what should become a medical tool, while the notification rate of 0.5% is low due to the average age of the participants. The greatest response (3.2%) was, in fact, in the over 65 s, confirming the increase in prevalence/incidence of arrhythmia with age and also how a study on AF should not be designed [7]. Moreover, the elderly use the devices to a lesser extent. Numerous other valid technological tools are also available to detect AF in this age group [3].

However, the purpose of the study was not AF screening, and therefore, the Apple Watch algorithm was designed to minimize false positives.

A positive AF finding in 34% of suspected cases is not relevant in itself, but the authors report that these subjects had a relatively high AF burden, with the majority of episodes lasting over an hour. Furthermore, AF often presents paroxysmal and infrequent patterns "AF can come and go," especially in the onset phases; therefore, it is plausible that it will not be detected in subsequent monitoring. The small number of people who returned EKG patches (20.8%), which were generally applied with a moderate delay, contributed significantly to the lack of identification of AF cases.

In summary, the Apple Heart Study highlights both the potential and limitations of digital innovation in medicine. The study can be considered preliminary for the design of studies conducted with more rigorous methodologies. In fact, the same researchers recognize its numerous limitations, for example, the "self-assessed" inclusion criteria (the participants themselves declared that they were not fibrillating, even if some later "confessed" a previous diagnosis of AF), the high number of dropouts after reporting, single-arm, open, self-reported data.

Reflections

The launch of the Apple Watch has stimulated the most discordant judgments between those who consider the device as opening the way to more predictive and preventive health care and those who instead classify it as a simple marketing operation for an expensive toy.

The Apple Watch warns users of the possible presence of an otherwise often difficult-to-identify pathology. In fact, AF is asymptomatic in over a third of cases, and its discovery can be completely random.

The increased engagement of citizens can also cause excessive confidence in "self-monitoring" and "do-it-yourself" diagnoses, even if, in the various explanatory screens to be read before using it, the app reminds us of the indispensability of medical examination to confirm and evaluate the presence of possible pathologies. The risk is that doctors will be overwhelmed by an enormous mass of information and new responsibilities in a context of greater uncertainty and confusion. It is, therefore, necessary to insert the device in predefined clinical paths, as well as automatically integrate the data into a single repository, for example, the patient's electronic medical record, not to further burden the bureaucratic burden of health-care professionals [8]. There is a need for expert doctors who educate the user on correct use, and inform about the possible risk of false positives and consequent false alarms in a population largely not digitally literate, are needed. Let us not forget that our identity is still fundamentally analog, even in an increasingly digital world.

The Apple Watch with EKG cannot replace traditional diagnostic tools. Its evolution from an expensive gadget to a reliable, effective, and efficient technology requires validation in research projects conducted with rigorous methodology. A goal could be greater knowledge of the natural history of AF, especially in asymptomatic cases, which cannot be easily diagnosed with traditional methods. The traditional treatment of arrhythmia is, in fact, based above all on data relating to the clinical presentation, and therefore, studies are needed to understand the populations that need to be screened and the cases of occult AF that actually require early treatment [9,10].

At present, there is no clear evidence that patients who are fibrillating at screening have the same risk as symptomatic and that anticoagulant treatment brings the same benefit. AI systems could therefore be used to identify patients who are most likely to have AF during screening and to benefit from related anticoagulation therapy, considering parameters such as the duration of the arrhythmia. In fact, it is not currently proven that screening improves clinical outcomes. Instead, it is safe that patients with clinical AF are under-treated [11].

Moreover, the use of the Apple Watch is not only a problem of effectiveness/efficiency but of cultural paradigm change. On the plus side, some watch companies have realized that such tools can be associated with medical devices rather than just gadgets. This entails the need to finance studies to evaluate the reliability and safety of the instruments, in close collaboration with universities and scientific societies, as in the Apple Heart Study, in analogy to what is usually done by pharmaceutical companies and manufacturers of traditional medical devices [12].

On the other hand, there is the risk that a sort of new sensory apparatus is being created, a pervasive instrumentation, capable of accessing physical, social, and environmental realities in ways, scales, and forms that have no precedent in the history of humanity. Data, digital projections of our people, and "one-way mirrors" can be fundamental inputs to produce advances in the medical field and improve health policies. Moreover, they are also tools for creating value in the digital market. The concept of personal and anonymous data has now disappeared in a sort of web of obsessive, out-of-control cataloging and profiling, in which the violation of privacy seems to be systematic and destined to accentuate with the imminent debut of the so-called "Internet of Things" (and of bodies). In this specific case, Apple reassures, stating that the data remains encrypted within the device, or iCloud, inaccessible to the company itself and made clear only by user intervention and recognition via code, touch, or face ID [13].

References

[1] Niola M. The present in a nutshell. Milan: Bompiani. 2016.

[2] Kotecha D, Breithardt G, Camm AJ, et al. Integrating new approaches to atrial fibrillation management. In: The 6th AFNET/EHRA consensus conference. Europace 2018;20:395–407.

[3] Zungsontiporn N, Link M. Newer technologies for detection of atrial fibrillation. BMJ 2018;363 k3946. Available from: <https://doi.org/10.1136/bmj.k3946>.

[4] <https://www.punto-informatico.it/apple-watch-ecg-italia-pro-contro/>.

[5] Perez MV, et al. Large-scale assessment of a smartwatch to identify atrial fibrillation. N Engl J Med 2019;381:1909–17.
[6] Campion EW. Watched by Apple. N Engl J Med 2019;381:1964–5.
[7] Colilla S, et al. Estimates of current and future incidence and prevalence of atrial fibrillation in the US adult population. Am J Cardiol 2013;112:1142–7.
[8] Sim I. Mobile devices and health. N Engl J Med 2019;381:956–68.
[9] Chen LY, Chung MK, Allen LA, et al. Atrial fibrillation burden: moving beyond atrial fibrillation as a binary entity: a scientific statement from the American Heart Association. Circulation 2018; 137: e623–e644. Available from: <https://doi.org/10.1161/CIR.0000000000000568>.
[10] Mandrola J, Foy A. Am Fam Physician 2019;99(6):354–55.
[11] Nicholas R, Jones NR, et al. Screening for atrial fibrillation: a call for evidence. Eur Heart J 2019;0:1–11.
[12] Santoro E. Agenda digitale.eu. Available from: <https://www.agendadigitale.eu/sanita/apple-watch-4-e-il-passaggio-da-gadget-a-dispositivo-medico-cosa-possiamo-imparare/>.
[13] Macri F. Available from: <https://francescomacri.wordpress.com/2019/03/28/salute-apple-watch-lorologio-fa-lelettrocardiogramma/.wordpress.com>.

SECTION 3

Digital intelligence and health

CHAPTER

9

Diagnostics and decision-making systems

Summary

The question that is systematically asked when evaluating the role of Artificial Intelligence (AI) in diagnostics is: "Has the superiority of AI over human specialists been demonstrated beyond any reasonable doubt?" The current answer is: not yet, at least in the vast majority of disciplines in which it has been properly tested.

As we will read in this chapter, in some specializations and for some well-defined applications (for example, in radiology), AI, and in particular deep learning devices, are, however, providing us with results of great interest.

In this chapter we will review the various specializations where AI has provided us with the most promising results.

In the medical field, practically any professional type is potentially involved in the use of AI, from prevention to assistance, from research to therapy. Moreover, the algorithmic approach can support above all those that E. Topol defines pattern doctors [1], that is, professionals who base their work on the interpretation of digital, radiological, retinal, histological, ophthalmic, dermatological, endoscopic images or those coming from various medical devices monitoring [2].

In particular, machine learning and deep learning (see glossary) not only process huge datasets of images but also of words, voices, videos, and digital and allow the recognition of *divergent pattern recognition* from normals; for example, alterations such as different densities, asymmetries, irregularities, etc., not perceptible with traditional diagnostic systems and which human experts themselves often fail to highlight, sometimes even to hypothesize a priori. The implementation of AI has the potential, still not fully expressed, to achieve faster and more effective diagnostics [3] and advanced predictive models to define the risk of certain diseases, as well as expand access to services that traditionally require specialized skills, such as example, as described below, screening for diabetic retinopathy in the primary care setting.

AI in Clinical Practice
DOI: https://doi.org/10.1016/B978-0-443-14054-9.00006-5

Another possible application is the search for new patterns, for example, of subtypes of neoplasms [4] and the possibility of optimizing decisions on specific treatments such as titration of fluids and vasopressors in the case of sepsis [5].

Such associations are the product of mathematical algorithms, and therefore, may, however, not have clinical significance or relevance. Furthermore, the association between some patient characteristics and treatment outcomes are correlations, not causal relationships. The results are, therefore, not appropriate for direct translation to clinical action but rather can serve as a research hypothesis to be evaluated in trials and other types of studies to establish the effective presence of a cause-effect relationship.

The superiority of AI systems in diagnostics has not yet been demonstrated with certainty. In the first systematic review and meta-analysis comparing deep learning systems and healthcare professionals, based on 31,587 studies, of which only 82 (out of 147 patient cohorts) were included, a substantial equivalence of performance between AI and clinicians was highlighted [6]. The study also found that only a limited number of papers had adequate external methodological validity and that the comparison groups, in particular, were often different.

Additionally, deep learning studies have shown poor reliability in data presentation, limiting the validity of the reported diagnostic accuracy. According to the authors, new reporting standards, which address the specific challenges of deep learning, could improve the reliability of future studies for this undoubtedly promising technology.

A team from Google Health offered a 3-step guide on how to critically evaluate research on AI, ML, and DL, underlining the need to combine traditional criteria with some specific considerations. For those wishing to learn more, please refer to the bibliographic reference [7]; here, it can be noted that the authors consider it essential to evaluate the quality of the algorithm training data. What characteristics of patients are measured and how? With easily implementable methodologies or with emerging technologies available only in experimental contexts?

It is cited that some of the ophthalmic diagnostic systems are trained on generic images, observed on the internet, of objects and places, and then used for other uses, such as retinal images. Therefore, even if they show good overall performance, they risk incurring systematic errors to the detriment of specific subgroups of patients.

It is also underlined that the authors of the articles on ML, in great development, should explain the premises, the properties, the optimization strategies, and the limitations to make the methodology and the interpretation of the results more understandable to readers (and reviewers) in nonspecialists in digital technologies.

Experts reiterate the importance of the study population, which should be representative of large communities, also recalling that a good performance only in a group of subjects can reflect the specificities of the training parameters rather than the difficulties in generalizing. The validation of the systems must take place in daily practice, on large and heterogeneous populations, on important clinical outcomes, such as mortality or morbidity, and not on simple "surrogate endpoints" (for example, the lowering of cholesterol as an indicator of improved health status).

Here we describe some of the countless applications in the medical field without claiming to be exhaustive but only to provide an overview [8].

Applications in the radiological field

The radiological field is the one most affected by the digital revolution. For example, 2 billion chest X-rays are taken worldwide each year. By typing the term "Artificial Intelligence" on PubMed, 99,526 publications have been obtained to date. Combined with the term "Radiology," there are 6937 of them, about 7%.

Although with many limitations, for example, the type of studies, generally retrospective, and above all, the lack of data confirmed in daily practice, the results in the experimental context are remarkable. In many cases, intelligent decision support systems[1] have demonstrated superior performance to that of professionals, so much so that, according to some observers, such as scientist Andrew Ng, specialists could be replaced by machines in a short time and discipline, from the point of view of humans, would be in extinction.

Certainly, radiologists, who use the visual pattern recognition system for their work, similar to Kahneman's thought system 1, are subject to possible bias, for example, the so-called "inattentional blindness," for which subjects, focused on specific aspects, they can overlook unexpected data even if perfectly visible. This was shown in a study in which 83% of radiologists did not see the image of the gorilla on the X-rays they were analyzing [9].

In general, a false positive rate of 2% and false negatives over 25% are calculated. In addition, some images may not be accurately evaluated by the human eye, such as texture, contrast, and signal strength.

1 Computerized programs that process clinical and scientific data to support the decisions of health professionals

Furthermore, the human interpretation of images is generally highly subjective and characterized by high intraindividual and interindividual variability.

AI obtains important results; for example, an AI algorithm has proved superior to a comparison group represented by four radiologists in pneumonia diagnostics, even if the level of accuracy (0.76 on the ROC curve, see box) it is optimal and in any case, the system is able to offer an answer to only a small part of the total daily work of radiologists [10]. In addition, for greater validation, a comparison with a larger number of specialists should be made. A team of Google researchers developed an algorithm that, analyzing the same images of the aforementioned study, made 14 different diagnoses, with excellent results, especially in the evaluation of heart enlargement and lung collapse (0.87) [11], conditions, which generally do not cause particular difficulties in the usual diagnostics!

In a study by Pranav Rajpurkar et al. with a deep learning algorithm called "CheXNeXt," trained on a dataset of over 100,000 chest radiographs of about 31,000 patients, results were obtained similar to the results by a group of nine radiologists in the diagnosis of pneumonia, pleural effusions, lung masses, pneumothorax, and nodules on simple anteroposterior chest X-rays. With the same diagnostic efficacy, the average interpretation time of the 420 images of the validation set was 420 minutes for humans and only 1.5 minutes for the algorithm [12]. This tool, which is still under development, will be able to offer solutions to the diagnostic errors of professionals due to lack of time or fatigue and provide the possibility of valid diagnostics in countries where radiologists are few in number or absent [13].

In another retrospective study, an AI system based on neural networks showed good results in the interpretation of neoplastic pulmonary nodules in the chest X-ray [14].

Also, in the pulmonary field, a study of over 470,000 reports demonstrated an acceptable performance in automatic triage on chest radiographs, classifying them as critical, urgent, nonurgent, or normal. Sensitivity was 71%, specificity 95%, positive predictive value 73%, and negative predictive value 94%. Reporting times were significantly reduced [15].

In a retrospective study, an automated deep learning system was found to be able to diagnose a standard chest X-ray pneumothorax with low sensitivity (0.80) but high specificity (0.90) for moderate and severe forms. The model could integrate the activity of radiologists by acting as an additive alarm system, especially in contexts with a high-performance flow, to exclude the most important types [16].

The AI Infervision system, which would allow a more accurate visualization of lung CT scans for COVID-19 pneumonia, in terms of volume, shape, and density, with a large reduction in time (10 seconds instead of the standard 15 minutes), did not currently produce articles in peer-reviewed journals. AI-guided lung CT scans could be particularly useful in contexts where there are too many cases of disease to perform all the necessary tests and the time to read the images normally is limited [17].

A document from the European Society of Radiology proposes to adapt Asimov's three rules of robotics [2] to AI applied to the discipline. In particular, AI systems should be very effective in diagnostics and always supervised by a radiologist, complementing human diagnostics and not a replacement. Furthermore, algorithms must be protected from possible obsolescence, due to the development of more modern machines or due to changes in clinical practice. The document also recognizes that Asimov's laws are fictitious and calls for institutional regulation, which ensures the application of certain and shared rules [18].

Certainly, AI systems can be used to train young radiologists and to improve intraindividual and interindividual reproducibility among experts.

The diagnosis of respiratory diseases: some commented paradigmatic studies

A few months ago, JAMA published an important study by South Korean researchers on the application of an artificial intelligence system such as *Deep Learning-Assisted Detecting* (DLAD) to radiological diagnostics [19]. In this interesting work, researchers have developed a diagnostic algorithm capable of recognizing four important groups of respiratory diseases on the basis of simple standard chest radiography: malignant neoplasms, pneumonia, active tuberculosis, and pneumothorax [20,21]. The researchers used 54,221 normal radiographs of 47,917 individuals and 35,613 pathological images of 14,102 individuals as a training base for the AI system. The interpretative algorithm created on the basis of these data was then tested on 486 normal and 529 pathological chest radiographs from five different diagnostic centers; the same radiographs were also interpreted by three groups of doctors, five internists, five generalist radiologists, and five thoracic radiologists for a total of 15 experts.

2 These are rigid principles, not to be transgressed, theorized to reassure humanity about the good "intentions" of robots. First Law: "A robot cannot harm a human being, nor can it allow a human being to receive harm due to its lack of intervention." Second law: "A robot must obey orders given by human beings as long as these orders do not contravene the first law." Third law: "A robot must protect its existence as long as this does not conflict with the first and second laws."

The diagnostic algorithm of the DL AD artificial intelligence system demonstrated better diagnostic accuracy than all comparator medical groups, including thoracic radiologists. More precisely, the DLAD proved superior to experts both in the identification of pathological X-rays (983 corrected out of 1000, while humans oscillated between 814 and 932 with $P < .05$), and in the correct localization of lesions (985 out of 1000 in favor of the DLAD while humans ranged between 781 and 907 with $P < .01$) (1).

As expected, thoracic radiologists provided the best diagnostic interpretations while the internists the least frequently correct ones; however, when the three groups of humans used the DLAD, their performance improved significantly without reaching the standards of artificial intelligence.

The study by Wang and colleagues is one of the most important in the field of radiological diagnostics using artificial intelligence systems both for the size of the sample used (just under 90,000 X-rays) and for the rigor of the checks carried out: the diagnostic system based on intelligence artificial was validated on about 1000 radiological findings provided by five different institutions, and examined by three different groups of professionals who carried out the human radiological diagnosis.

The results are clear and unambiguous; the DLAD artificial intelligence diagnostic system has demonstrated a high accuracy in identifying even small lung lesions, with a good positive predictive value and an excellent negative predictive value. In this first diagnostic level, DLAD has proven superior to all human specialist groups, not only internists but also thoracic radiologists.

An evident limitation in the discussion is a lack of precision of the DLAD in the differential diagnostics between the various pictures, in particular, between pneumonia and tuberculous forms, where more frequently, man, perhaps using qualitative assessments, seems to be more precise. An intrinsic limitation of the method, as correctly stated by the authors, is linked to the fact that it was developed in a particular context (South Korea), which, although technologically very advanced, could provide indications that cannot be transferred to other contexts.

An indisputable result of the study is the demonstration that systems such as the DLAD are, however, extremely precious tools to support human diagnostics, both as a first level of diagnosis, with the function of "superassistant," and as an important tool for in-depth analysis in the most complex situations.

Conclusions

Digital diagnostics advances with surprising speed and precision, and in some respects, disconcerting. Nations once considered marginal, such as South Korea or Taiwan, invest great economic and human resources in these sectors and humbly and silently are building not only theirs but also our future.

It is good to remember that also in this field, our beloved Italy boasts an enormous and enviable cultural and scientific tradition; let us not forget that the very first prototype of the personal computer was developed in Italy in 1962 by Eng. Perotto of Olivetti and that the microprocessor was invented in 1968 by Federico Faggin (see bibliographic study).

We are very worried about the myopia of a large part of our ruling class, which today, even more than earlier, does not dedicate adequate economic and intellectual investments to the development of digital culture in our country.

The simple chest X-ray as a predictor of mortality

A recent study by some radiologists from Harvard Medical School used Neural Networks and Deep Learning processes and obtained new interesting information from the banal chest X-ray [22]. The knowledge of the process of creating digital images allows us to understand the great potential but also some potential limitations of the study.

In fact, if we remember that the digital system is based on numerical data, the interpretation of images made up of several "objects" in a reciprocal relationship (for example, large vessels of the heart (mediastinum)), which the human mind solves in the blink of an eye, forces the digital system to various attempts of trial, error, and error correction, with sometimes questionable results.

In the study in question, the researchers made reference to the "Transparent reporting of a multivariable prediction model for individual prognosis or diagnosis" (TRIPOD) [23] to catalog the chest X-rays of 41,856 patients and then developed a valid interpretative model that was subsequently tested on about 16,000 other patients to verify its effectiveness and usability. Basically, the study demonstrates with certainty that the simple chest X-ray, thanks to the neural network system used in the study, is able to insert the subject examined in one of the three main risk ranges; high risk, intermediate, low risk; the predictive value is high for the high-risk group both for deaths from lung cancer and for deaths due to heart problems or respiratory problems; the negative predictive value for the low-risk groups was also satisfactory, while in the intermediate risk classes, in the opinion of the authors themselves, additional risk factors, not detected by the neural network in simple radiographs, should be taken into consideration. Therefore, according to the researchers, the results would be comforting and would offer radiologists and clinicians new tools to formulate prognostic hypotheses, thus identifying early and appropriately treating subjects at risk.

Comment

The study is important not so much for the results, which the authors themselves believe can be improved, as for having demonstrated on tens of thousands of patients that the simple standard anteroposterior chest X-ray is able to provide much more information and predictions than most of us could suspect.

A traditional study on samples of that large number would probably have provided even more precise and in-depth information but with enormously greater use of human, economic, and time resources than those used.

Studies such as the one presented are, therefore, certainly very useful just as pilot studies; in particular, it will be very interesting to compare the type of prognostic classification that emerges from this study with other classifications based on data that can be found with equal ease and to evaluate whether the survival forecast based on chest X-rays really provide useful data in clinical practice, or if in the end, it does not limit itself to repurposing, albeit in "mathematically correct" terms, the old, wise, and witty judgment of the expert clinician who, faced with a "bad plate," he comments "I really don't like this slab, I think this poor man will have a short life…." The old clinician wasn't sure why but he was worried, and most of the time, unfortunately, he was right.

Even better performances were achieved in diagnosing wrist fractures in emergency settings with sensitivity improvements (81% to 92%) and a 47% reduction in misinterpretations [24].

Other applications of AI systems have concerned a wide range of radiological images: bone fractures, aging estimation, classification of tuberculosis, vertebral compression fractures, CT scan for pulmonary nodules, liver masses, pancreatic neoplasms, coronary *calcium score*, hemorrhages, brain trauma, MRI images, echocardiograms, and mammograms. These studies, however, offer very different results depending on the pathologies, and the same methods of evaluating the results are not necessarily the best for clinical use. Many studies have also not been published in peer-reviewed journals.

In particular, in the context of acute neurological events such as stroke and head trauma, important results have been obtained in terms of speed in diagnostic interpretation. Moreover, the accuracy was inferior to that of humans … the road is still long!

Breast cancer diagnostics

One of the areas of study of AI systems is **mammography screening**, which aims to diagnose breast cancers at an early stage, in the absence of obvious signs of disease, when treatment can obtain the best results. The interpretation of mammograms is also characterized by possible false positives and false negatives. The physiological mammary radiological density can, in fact, mask the tumor. Mistakes by examiners are also inevitable. This has developed a great interest in the creation of AI systems capable of improving the diagnostic performance of radiologists [25].

An example is the AI system developed by Google, which, according to a study published in the authoritative journal Nature, would be able to reduce false positives by 5.7% and 1.2% and false negatives by 9.4% and 2.7%, using US and UK databases, respectively. The AI system provided superior performance both to the historical reports provided previously on the same databases and to those six radiologists who interpreted 500 randomly selected radiological images in a controlled study [26]. A commentary editorial also warns against possible enthusiasm: the real world is more complex than the "ideal and unreal" world of research [27]. For example, in the study, most of the images were taken by the same machine, and the results that could have been obtained with other mammography devices are not known. It would also be important to know the performance of AI with respect to the two different types of mammography used, tomosynthesis (3D mammography) and conventional digital (2D), characterized by different performances. Finally, in the study, the characteristics of the population are not well defined, apart from age, which are essential for the generalizability and applicability of the technology.

The high hopes in computer-assisted breast cancer diagnostics, raised by experimental studies and the availability of large databases for training ML algorithms, have not been confirmed by "real world" studies [28]. In particular, a worsening of sensitivity was noted, that is, the ability of radiologists to highlight the presence of the neoplasm, with an increase in false negatives, however, without improving the specificity, that is, the ability of specialists to exclude the presence of the neoplasm and therefore increasing the false positives.

A study by M. Bahl used an ML model, which, based on more than 1000 mammograms followed by biopsy at high risk of cancer, showed that more than 30% of subsequent surgeries could have been avoided [29]. A large database analysis by the *US Breast Cancer Surveillance Consortium Registry* also found no improvement in diagnostic accuracy with AI algorithm technology [30]. A possible cause of poor efficacy in practice would be the fact that radiologists ignored or did not use algorithmic information due to the high frequency of later unconfirmed pathological reports. Another reason could be that the AI training was carried out using previous "human" diagnoses, sometimes characterized by false negatives, that is, failure to report neoplasia in images that are actually pathological. In general, an alteration reported by AI, even if not recognized as such by radiologists, should require further analysis, although this may lead to an increase in false alarms for patients.

Performance monitoring is fundamental to "teach" algorithms. According to Etta D. Pisano, a great help can come from the data of the electronic folders, as long as privacy, security, and reasons for use are safeguarded [27]. Further studies are certainly needed to define the real role to be assigned to AI systems in the diagnosis of breast cancer.

Applications in oncology: Watson for oncology

IBM, starting with IBM Watson, whose name does not derive from that of the famous medical collaborator of the character of Sherlock Holmes but from the founder of the same computer industry, Thomas J Watson, has launched several projects in the medical field such as Watson for Genomics, Watson for Drug Discovery, Watson for Clinical Trial Matching. In the oncology field, he created Watson for Oncology (WFO), one of the best-known cases of the use of artificial intelligence in medicine, an evolution of traditional decision support systems. It is an expert system trained by the oncologists of the Memorial Sloan Kettering Cancer Center in New York through the administration of therapeutic protocols and rules to learn how to apply them in different cases. Used, for a fee, in dozens of hospital centers around the world, from South Korea to Slovakia, from India to Florida, it supports its clinical activity by comparing patient data with the literature produced by medical journals, international guidelines, and with the "historical" similar real cases faced in the past, to propose the most appropriate treatment, taking into account the effectiveness of the therapies and side effects [31].

However, uncertainties and limitations of the system are described [32]:

- algorithms based on relatively small numbers with very limited real-world data
- possible methodological biases: the data used to train the system, and also the therapeutic protocols implemented, used as standards, come from the clinical histories of American patients and are mainly based on US studies and guidelines, which can lead to reproducibility problems in other populations and contexts, including economic ones.
- limited number of tumor types that the system is able to recognize
- difficulty reinstructing the system whenever the guidelines and studies on which its decisions are based completely change or are otherwise updated
- lack of randomized clinical trials published in peer-reviewed journals that demonstrate their reliability and/or greater efficacy on clinical outcomes compared to traditional systems

- reservations on the protection of the privacy and safety of citizens and patients
- system regulation problems and definition of specific responsibilities in case of error and allegations of malpractice (for example, the system recommended treatment with a drug in a patient with severe bleeding, a typical contraindication of the same)
- difficulties in integrating doctors and nurses into the working context, with different levels of use in different hospitals
- black box effect of algorithms; lack of evidence underlying their "reasoning," such as to allow doctors to evaluate whether to decide to follow the suggestion or not
- danger of overreliance and excessive dependence on such systems that could have serious effects of de-qualification and desensitization of doctors to the clinical context.
- ethical issues, such as using an artificial intelligence system to make end-of-life care decisions; or great perplexity in case of dialog between two artificial intelligence systems in a language unknown to man [33]

AI systems, through radiological indices and sophisticated algorithms, have been proposed to evaluate the effects of antineoplastic therapies, in particular immunotherapy. It is believed that with AI applied to radiology, radiomics will be able to provide more precise information on the volumetric and morphological changes of primary neoplasms and/or single metastatic lesions. Radiomic analysis can, in fact, extract hundreds of features from CT, MRI, and PET beyond human capabilities (*texture analysis*), and correlate them with other data (proteomics, genomics, liquid biopsies, etc.) creating accurate patient profiling useful for prediction of clinical outcomes and response to treatments [25,34].

In the oncology field, the investigation coordinated by Eugenio Santoro, Head of the Medical Informatics Laboratory, and carried out by the Mario Negri Institute of Pharmacological Research in Milan on 537 patients in collaboration with AIMAC (Italian Association of Cancer Patients, relatives, and friends) [35]. The conclusions are that 74% of patients use the internet as an important form of obtaining information (after the oncologist and before the general practitioner). In particular, the most used search engine is Google (62%), but more institutional sites, such as those of scientific societies, health institutions, and patient associations, are also consulted (about 40%). Social media platforms are rarely used, with the sole exception of *online communities*,

continued

(*cont'd*)

which are useful for exchanging information between "peers" but are generally considered unreliable by 51% of the interviewees. Instead, doctors and oncologists have little propensity to suggest websites and portals for patients.

32% of cancer patients use at least one health app for smartphones, mainly to use health services, such as bookings for medical visits or exams and access to reports and to monitor physical activity. The apps dedicated to food follow.

Apps for monitoring health parameters and improving adherence to care were not widely used. It is interesting that 8 out of 10 patients among those who do not use apps or wearable devices would be willing to do so if the doctor or oncologist suggested them.

Finally, cancer patients make extensive use of contact tools with the doctor, especially email and WhatsApp, to follow the sms.

AI systems based on *natural language processing* (see glossary) can analyze data on clinical outcomes reported by patients in a more precise and personalized way than with traditional questionnaires. Furthermore, the data are preliminary and require further checks before being placed on the market.

Applications in the anatomopathological field

Pathologists have adopted AI systems for reading histological slides with greater latency, due to various technical problems, mainly due to the need to convert standard images into the latest digital formats. Furthermore, these professionals also suffer from difficulties in sharing diagnoses, for example, due to the small quantity of the tissue sample, the result of which is the heterogeneity of interpretation. For example, in some forms of breast cancer, the consensus among pathologists may be less than 48%. As for other professionals, there is also the time factor available to draw up the reports. Deep learning offers the potential to improve the accuracy and especially the speed of diagnostics, as highlighted in some retrospective studies [36].

Some studies have evaluated the possibility of AI to classify neoplasms, specifically mammary and pulmonary, without the direct participation of pathologists. By using tumor DNA methylation patterns, classifications rarely used in the clinic can be obtained. Deep learning algorithms can finally highlight genomic mutations that cannot be detected with common diagnostic means. The value of these technologies must, however, be tested in validation studies on clinical endpoints.

The first prospective study of AI in a real setting highlighted the improvement in diagnostic accuracy obtained from the synergy between algorithms and doctors. The list of FDA-approved algorithms for image interpretation is constantly expanding, although only a limited number have published the results in peer-reviewed journals.

The limits of AI in cytological and histopathological diagnostics

Initially, a set of images is required that pathologists have classified, for example, as "cancer" or "noncancer." Using a subset of such images (the *training set*), the computer, based on the renaissance of patterns (color, shape, edges, etc.) learns to discriminate between malignant and benign tissues without explicit programming. The performance of the algorithm is evaluated on the further remaining images, which the machines have not previously analyzed (the *test set*). Eventually, the algorithm is further tested on further images, and its diagnostic capability is further refined. At each step, the validity of the algorithm depends on the quality of the initial interpretative standard (*external standard*).

The fundamental problem is the lack of a histopathological gold standard, above all because of the histopathological classification being static and photographic, the morphology of the single cells, the surrounding tissue architecture, the relationship between these parameters and various biomarkers. The fundamental information for the clinician is instead dynamic, on the evolution of the neoplasm, by metastasis or local infiltration. Diagnosis, in particular, of the early stages of cancer leads to discordant evaluations and often to overdiagnosis. Unfortunately, even ML, which certainly allows more constant and reproducible diagnoses, does not solve these problems. According to A. Adamson and Gilbert Welch, it could even accentuate the overdiagnosis. Their proposal would be to use, in the initial phase of diagnostic interpretation, different groups of pathologists in order to allow the algorithms to discriminate different diagnostic interpretations, total agreement on cancer, total agreement on absence, especially disagreement on the presence or absence of neoplastic cells, highlighting the so-called "gray zone." This would allow pathologists to focus on more ambiguous findings to consider more conservative treatments for lesions of more uncertain significance. Finally, it could foster further studies and research on the natural history of such injuries [3].

Applications in the dermatological field

The accuracy of AI algorithms for diagnosing skin cancers (epitheliomas, basaliomas, and melanomas) has been demonstrated in many studies comparing dermatologists; sometimes AI has shown greater accuracy, for example, in a study by Andre Esteva, using an algorithmic deep learning technique on nearly 130,000 skin lesion images and over 2000 diagnostic reports; the AUC was 0.96 for skin carcinomas and 0.94 for melanomas against 21 dermatologists [37]. Effects on general population series, where patient risk profiles are different, are to be demonstrated [38].

The use of such AI systems could be particularly useful in primary care to integrate the GP's expertise with a support technology capable of providing a specialist-level service, for example, through the app of a connected smartphone to an advanced and validated technology [39]. Moreover, no studies have been carried out in a clinical setting but only experimentally. Several apps for the diagnosis of skin cancer have not shown better performance than dermatologists, especially for false negatives. Their expanding market requires strict regulation and rigorous clinical comparison studies [40].

Applications in ophthalmology

Numerous studies have compared the performance of AI with that of ophthalmologists, especially in diabetic retinopathies and age-related macular degeneration. In a paper, deep learning training was performed on a huge database of over 128,000 retinal photographs evaluated by 54 ophthalmologists, while the clinical evaluation involved over 10,000 retinal images of 5000 patients with diabetic retinopathy. The comparison compared the DL with 7/8 specialists in two different assessment settings. The result, an AUC of 0.99, was overlapping and optimal in terms of accuracy [41].

A prospective trial conducted in 10 primary care centers, using a deep learning algorithm on 900 diabetics with no known retinopathy, allowed the Food and Drug Administration (FDA) in 2018 to certify the first AI-based medical device [42]. The software, called "IDx-Dr," uses an algorithm to analyze retinal images obtained from a camera and is able to detect a higher than mild level of diabetic retinopathy in the absence of medical support with a sensitivity of 87% and specificity of 90% [43]. The use of the system by nonspecialized personnel working in primary care centers can allow the early diagnosis of cases of probable disease and, depending on the stage of retinopathy, recommend specialist advice or check-up after 12 months. The study is a milestone, as it is the first evaluation of an AI in the clinical setting, even if the results are not as high as in the experimental setting.

A similar prospective study, conducted using deep neural networks in a real-world context, with exams carried out by nurses, instead showed a high rate of false positives, probably due to the reduced incidence of retinopathy and the low quality of the images obtained in real conditions [44].

A study conducted using OCT allowed the patient to observe a video that showed the portions of the scans that were used for the conclusions of the algorithm accompanied by the level of confidence of the diagnosis: a first attempt to make the so-called "AI black boxes" transparent.

Another area of development, still in a highly experimental phase, concerns the use of ocular images for the early diagnosis of dementia, including Alzheimer's and cardiovascular predictivity.

Other eye diseases diagnosed by neural networks are early-stage glaucoma, congenital cataract, and retinopathy in premature babies.

Track AI, a software based on AI and connected to smartphones, capable of identifying visual disturbances in children by analyzing the pathological movements of the pupil, is being developed.

A Chinese study used ocular data from about 130,000 young people aged 6 to 20 from computer files in eight ophthalmic centers, and demonstrated a good ability to identify subjects at high risk of myopia. The model is used in prospective studies to determine whether behavioral or clinical interventions can delay progression in high-risk children.

Applications in the cardiology field

The diagnostics of electrocardiograms with automatic reading dates back to many years, however, with poor results. The use of neural networks has shown excellent sensitivity (93%) and specificity (90%) in the diagnosis of acute coronary syndromes, comparable to those of cardiologists [45]. Similar good results have been found in diagnosing arrhythmias with single-lead EKGs [46].

An important use of IDs is the recognition of undiagnosed AF, although clinical evaluation trials are limited. Moreover, the data on which the treatment of AF is based are those relating to the clinical presentation, and therefore, studies are needed to understand the populations to be screened and the cases of occult AF that actually require treatment.

In the field of echocardiography, the algorithms have provided better results than groups of expert cardiologists, but only in the experimental [47] setting. AI systems are also able to quantify other parameters, such as vascular stenosis [48] and coronary ischemia, avoiding invasive measures [49].

In a retrospective study, AI demonstrated great accuracy in diagnosing hypertrophic cardiomyopathy, cardiac amyloidosis, and pulmonary hypertension [50].

There are also smart stethoscopes, which can analyze heart and lung sounds for personal patient profiling. In the future, it is possible that cardiologists will be able to grasp information hidden from the human senses but capable of predicting cardiac events [51].

Decision support systems would have identified new predictive models of cardiovascular risk capable of improving prognostic accuracy beyond that of traditional methods [52].

In a study conducted by the research group of Ziad Obermeyer in Boston, the subjects who arrived at the Brigham and Women's Hospital emergency room, based on the risk of myocardial infarction predicted by a machine learning algorithm, were divided into 10 groups for each of which the number of tests performed (stress test or cardiac catheterization, within one week of the visit) and the number of revascularizations performed (within one week of the test) were calculated. For subjects included in the low-risk decile, the prescription of diagnostic tests was associated with a clinical advantage in only 1.7% of cases. Translating these data into cost-effectiveness, the author calculates that in the low-risk group, these tests are associated with an annual cost per life saved of nearly $600,000, well above the $100,000 threshold typically used to define an acceptable ratio.

The scholar also analyzed the variability among doctors in terms of risk assessment: it was found that doctors who prescribe the largest number of tests do so up to three times more with patients at low risk (negative data) but also up to two times more with those at high risk (positive data). It is, therefore, about finding the right compromise between these two trends, and AI can help for a greater understanding of doctors' behaviors. In this case, the technological applications have made it possible to understand that the risk assessment process of patients with suspected symptoms of myocardial infarction determines a negative cost-effectiveness ratio [53].

In a retrospective cohort study of over 9000 asymptomatic patients (mean age 57.1 years, 56% women, 44% men), who underwent abdominal CT scans for colorectal cancer screening, the authors found that five biomarkers from AI algorithms (subcutaneous abdominal fat, muscle, hepatic and bone density, aortic calcification), were more predictive of cardiovascular events, such as myocardial infarction, stroke, and death than the classical Framingham score and the BMI (body mass index) [54]. The study opens up new and fascinating predictive possibilities, even if there are numerous limits to its practical application, for example, the lack of evaluation of the risk-benefit ratio on clinical endpoints and of a sensitivity and specificity analysis, the need for expertise for the execution of exams, additional costs [55].

Applications in the gastroenterological field

Excellent results have been obtained in the diagnosis of very small intestinal adenomas (<5 mm) and sessile polyps. In the first prospective validation study of IA, on 325 patients with 466 small polyps, the accuracy at routine colonoscopy was 94% with a negative predictive value of 96% and a diagnosis time of 35 seconds [56]. These results were replicated in another study [57].

ML programs were applied to identify patients not compliant with common screening programs but at high risk of colorectal cancer, using the Colon-Flag Test, an algorithm that integrates age, sex, and a simple blood count [58].

Another possible use of deep learning is in the classification of liver masses, in the staging of fibrosis, and in guiding doctors in performing liver biopsies.

Applications in the neurological field

Several papers have evaluated emerging technologies for long-term EEG monitoring in epileptic patients, in various contexts, with encouraging results, especially for possible use in a home setting [59]. The EEG detection was also used to quantify, through specific algorithms, the sensation of pain, obtaining useful information, for example, to objectify the duration of action of analgesics. Studies are underway to compare this information with that of common subjective scales [60].

In Parkinson's, motor disorders at an advanced stage have also been studied, such as impaired gait, *freezing*, and balance disorders, especially in situations that are generally difficult to explore, for example, during the night. This may be useful in assessing the effectiveness of the therapy more accurately.

Deep learning can be very useful for highlighting cerebral aneurysms at angioRM and quickly distinguishing ischemic from hemorrhagic stroke, localization of any bleeding source, and extension of the lesion, reducing possible diagnostic, and therefore, therapeutic delays.

Applications at the level of health systems

ML-based decision support systems have been tested in the hospital setting, for example, to estimate the probability of readmission after discharge, the risk of complications, such as sepsis and septic shock, and postsurgical complications, to optimize heparin therapy, in the prediction of the risk of [61] venous thromboembolism, to estimate the effective benefit of palliative care in individual cases.

Machine vision systems have been used in hospitals to monitor activities, such as hand washing, by healthcare workers, the status of critically ill patients in intensive care units, and the risk of patients falling.

In general surgery, ocular microsurgery, radiotherapy, and artificial vision method have been used. Furthermore, further studies are needed to evaluate safety, efficacy, and the possibility of cost reduction.

The use of AI systems for image interpretation and decision support to doctors could allow for the reduction of the workload and also for administrative staff, for example, by reducing the staff required for code implementation, reporting, and booking appointments.

It is possible to imagine a future characterized by AI-based **virtual assistants** supporting both doctors and patients. Healthcare professionals can already use chatbots that[3] provide up-to-date information on the scientific literature. Numerous initiatives are available for patients, not all validated, to provide answers to the most common questions of patients, for example, oncology, or to improve specific functions such as the cognitive activity of patients with Alzheimer's disease.

Machine learning systems can contribute to the reduction of medical errors, often attributed to cognitive bias, favored by work overload and fatigue. The algorithms are able to report the probability of errors during medical activity in a given context, such as the MedAware software, which provides, for drug prescriptions, *alerts* that a monitoring analysis has shown valid in ¾ of cases and overall clinically useful. The authors propose caution because the validity of the system is highly dependent on the quality and completeness of the data entered [62].

AI algorithms have been used in many other settings, for example, in anesthesiology to avoid periods of hypoxia during surgery, in the design of trials in oncology, in the personalization of radiotherapy treatments, in the stratification by the severity of prostatic neoplasms on MRI, in the diagnosis of knee injuries with MRI and also in the selection of embryos for in vitro fertilization and in many other conditions;: **practically throughout the course of human life**.

3 software designed to simulate a conversation with a human

In general, machine learning can contribute to the improvement of health policies at the population level, elaborating huge datasets that integrate medical images, electronic records, and genetic data, creating models that allow to carry out interventions, especially in the hospital setting, on high-risk subjects hospital mortality, rehospitalization within 30 days of discharge, excessive length of hospital stay.

The data can also be obtained from other sources, for example, from social media, such as Twitter, to identify the illegal sale of opiates [63].

Applications in the genetic field

Machine and deep learning algorithms allow the analysis of datasets of genomic biology and in general -omics (see Table 9.1). A current application with great potential for development is precision medicine.

Important results have been obtained in the classification and analysis of genomic variants, neoplastic somatic mutations, gene-gene interactions, and RNA sequencing in the analysis of the microbiome, allowing an approach not only at the level of a single dimension omics but also with multiomics algorithms able to integrate the different datasets.

In the future, one could think of using a large amount of data of each individual (anatomical, biological, physiological, environmental, socioeconomic, and behavioral), together with treatments carried out and outcomes, to build a huge personal database and create a "digital twin," equivalent to those used in aeronautics, to simulate on a virtual level preventive interventions, treatments, and outcomes in various conditions [48].

TABLE 9.1 Mini glossary.

Genetics: science that studies the genes, heredity, and variability of organisms**Genomics:** study of the genome as a complete system, total amount of DNA of a cell or organism or species
Transcriptomics: study of the transcriptome, that is, of the set of messenger RNA molecules, to analyze the gene expression
Proteomics: study of the proteome, that is, the set of proteins of an organism or a biological system or the proteins produced by the genome
Metabolomics: study of the metabolome, that is, the set of all metabolites of a biological organism, end products of its gene expression, to detect its functioning (normal, reduced, or excessive)
Epigenomics: study of the epigenome, that is, heritable phenotypic changes in gene expression caused by mechanisms other than changes in the genomic sequence

References

[1] Topol E. Deep medicine: how artificial intelligence can make healthcare human again; 2019.
[2] Jiang F, Jiang Y, Zhi H, et al. Artificial intelligence in healthcare: past, present and future. Stroke Vasc Neurol 2017;2:e000101. Available from: https://doi.org/10.1136/svn-2017-000101.
[3] Adamson A, Welch HG. Machine learning and the cancer-diagnosis problem-no gold standard. N Engl J Med 2019;381:2285–7.
[4] Li A, et al. Unsupervised analysis of transcriptomic profiles reveals six glioma subtypes. Cancer Res 2009;69(5):2091–9. Available from: https://doi.org/10.1158/0008-5472.CAN-08-2100.
[5] Komorowski M, et al. The artificial intelligence clinician learns optimal treatment strategies for sepsis in intensive care. Nat Med 2018;24(11):1716–20. Available from: https://doi.org/10.1038/s41591-018-0213-5.
[6] Liu X, et al. A comparison of deep learning performance against health-care professionals in detecting diseases from medical imaging: a systematic review and meta-analysis. Lancet Digital Health 2019;1:e271–97 https: //doi.org/10.1016/S2589-7500(19)30123-2.
[7] Liu Y, et al. How to read arctic that use machine learning: users'guides to the medical literature. JAMA 2019;322:1806–16.
[8] Reddy S, et al. Artificial intelligence-enabled healthcare delivery. J R Soc Med 2019;112(1):22–8. Available from: https://doi.org/10.1177/0141076818815510.
[9] https://www.npr.org/sections/health-shots/2013/02/11/171409656/why-even-radiologists-can-miss-a-gorilla-hiding-in-plain-sight/
[10] Wang X. et al. ChestX-ray8: hospital-scale chest x-ray database and benchmarks on weakly-supervised classification and localization of common thorax disease. https://www.researchgate.net/publication/320068322_ChestX-ray14_Hospital-scale_Chest_X-ray_Database_and_Benchmarks_on_Weakly-Supervised_Classification_and_Localization_of_Common_Thorax_Diseases.
[11] Li Z. et al. Thoracic disease identification and localization with limited supervision. https://arxiv.org/pdf/1711.06373.pdf.
[12] Rajpurkar P, et al. Deep learning for chest radiograph diagnosis: a retrospective comparison of the CheXNeXt algorithm to practicing radiologists. PLoS Med 2018;15(11):e1002686. Available from: https://doi.org/10.1371/journal.pmed.1002686. eCollection. 2018. Nov.
[13] Saria S, et al. Better medicine through machine learning: what's real, and what's artificial? PLoS Med 2018;15(12):e1002721. Available from: https://doi.org/10.1371/journal.pmed.1002721.
[14] Nam JG, et al. Development and validation of deep learning-based automatic detection algorithm for malignant pulmonary nodules on chest radiographs. Radiology. https://pdfs.semanticscholar.org/e0ac/105018b58ed692412a8ef339b46866fa11f3.pdf?_ga = 2.250173958.1837388680.1582968967-1431564134.1576831057.
[15] Annarumma M, et al. Automated triaging of adult chest radiographs with deep artificial neural networks. Radiology 2019;291:196–202.
[16] Taylor Judy, et al. Automated detection of moderate and large pneumothorax on frontal chest x-rays using deep convolutional neural networks: a retrospective study. PLoS Med 2015;15(11):e1002697. Available from: https://doi.org/10.1371/journal.pmed.1002697.
[17] McCall B. COVID-19 and artificial intelligence: protecting health-care workers and curbing the spread Lancet Digital Health; 2020.

[18] Neri E, de Souza N, Brady A, et al. What the radiologist should know about artificial intelligence - an ESR white paper. Insights Imaging 2019;10:44. Available from: https://doi.org/10.1186/s13244-019-0738-2.
[19] Hwang EJ, Park S, Park CM, et al. Development and validation of a deep learning – based automated detection algorithm for major thoracic diseases on chest radiographs. JAMA Netw Open 2019;2(3):e191095. Available from: https://doi.org/10.1001/jamanetworkopen.2019.1095.
[20] McComb BL, Chung JH, Crabtree TD, et al. Expert panel on thoracic imaging. ACR appropriateness criteria routine chest radiography. J Thorac Imaging 2016;31(2): W13–15. Available from: https://doi.org/10.1097/RTI.0000000000000200.
[21] Speets AM, van der Graaf Y, Hoes AW, et al. Chest radiography in general practice: indications, diagnostic yield and consequences for patient management. Br J Gen Pract 2006;56(529):574–8.
[22] Lu MT, Ivanov A, et al. Deep learning to assess long-term mortality from chest radiographs. JAMA Netw Open 2019;2(7):e197416. Available from: https://doi.org/10.1001/jamanetworkopen.2019.7416.
[23] Transparent reporting of a multivariable prediction model for individual prognosis or diagnosis (TRIPOD): the TRIPOD statement. http://www.equator-network.org/reporting-guidelines/tripod-statement/.
[24] Lindsey R, et al. Deep neural network improves fracture detection by clinicians. Proc Natl Acad Sci USA 2018;115:11591–6.
[25] Neri E, de Souza N, Brady A, et al. What the radiologist should know about artificial intelligence - an ESR white paper. Insights Imaging 2019;10(44). Available from: https://doi.org/10.1186/s13244-019-0738.
[26] McKinney SM, et al. International evaluation of an AI system for breast cancer screening. Nature 2020;577:89–94.
[27] Pisano ED. AI shows promise for breast cancer screening. Nature 2020;577:35–6.
[28] Lehman CD, et al. JAMA Intern Med 2015;175:1828–37.
[29] Bahl M, et al. High-risk breast lesions: a machine learning model to predict pathologic upgrade and reduce unnecessary surgical excision. Radiology 2018;286(3):810–18.
[30] Kohli A, Jha S. Why CAD failed in mammography. J Am Coll Radiol 2018;15:535–7.
[31] Santoro E. Artificial intelligence, medicine, society. Res Pract 2018;34:27–8.
[32] Santoro E. Watson accompanies the work of oncologists; Forward 08.
[33] Field M. Facebook shuts down robots after they invent their own language. Telegr 2017; August 1.
[34] Vernuccio F, et al. Radiomics and artificial intelligence: new frontiers in medicine. Recent Prog Med 2020;111:130–5.
[35] https://www.marionegri.it/magazine/pazienti-oncologiche-strumenti-digitali.
[36] Wang, D., et al. Deep learning for identifying metastatic breast cancer. arX 2016.
[37] Esteva A, et al. Dermatologist-level classification of skin cancer with deep neural networks. Nature 2017;542:115–18.
[38] Challen R, et al. Artificial intelligence, bias and clinical safety. BMJ Qual Saf 2019;28:231–7. Available from: https://doi.org/10.1136/bmjqs-2018-008370.
[39] Leachman S, Merlino G. The final frontier in cancer diagnosis. Nature 2017;542 36-28.
[40] Digital oncology apps: revolution or evolution? The lancet oncology; 2018.
[41] Gulshan V, et al. Development and validation of a deep learning algorithm for detection of diabetic retinopathy in retinal fundus photographs. JAMA 2016;316(22):2402–10.
[42] https://www.fda.gov/news-events/press-announcements/fda-permits-marketing-artificial-intelligence-based-device-detect-certain-diabetes-related-eye.

[43] Abramoff MD, et al. Pivotal trial or fan autonomous AI-based diagnostic system for detection of diabetic retinopathy in primary care offices. NPJ Digit Med 2018;1(39). Available from: https://doi.org/10.1038/s41746-018-0040-6.eCollection.2018.
[44] Kanagasingam Y, et al. Evaluation of artificial intelligence-based grading of diabetic retinopathy in primary care. JAMA Netw Open 2018;1(5):e182665. Available from: https://doi.org/10.1001/jamanetworkopen. 2665.
[45] Strodthoff N, et al. Detecting and interpreting myocardial infarction using fully convolutional neural networks. Physiol Meas 2019;. Available from: https://iopscience.iop.org/article/10.1088/1361-6579/aaf34d.
[46] Rajpurkar P. et al. Cardiologis-level arrhythmia detection with convolutional neural networks arXiv: 1707.01836 [cs.CV].
[47] Madani A, et al. Fast and accurate view classification of echocardiograms using deeplearning. NPJ Digit 2018;1:6.
[48] He J, et al. The practical implementation of artificial intelligence technologies in medicine. Nat Med 2019;25:30–6.
[49] Hae H, et al. Machine learning assessment of myocardial ischemia using angiography: development an retrospective validation. PLoS Med 2018;15(11):e1002693. Available from: https://doi.org/10.1371/journal.pmed.1002693.
[50] Zhang, et al. Fully automated echocardiogram interpretation in clinical practice feasibility and diagnostic accuracy. Circulation 2018;138:1623–35.
[51] The heart of the matter: technology in the future of cardiology. the medical futurist; 2019.
[52] Weng SF, Reps J, Kai J, Garibaldi JM, Qureshi N. Can machine-learning improve cardiovascular risk prediction using routine clinical data. PLoS One 2017;12(4):e0174944. Available from: https://doi.org/10.1371/journal.pone.0174944.
[53] Obermeyer Z. Artificial intelligence at the service of medical decisions. In: 4words. The words of innovation in healthcare Supplement to Recent Advances in Medicine 2018; Vol 109, Iss 4, April 2018.
[54] Pickhardt PJ, et al. Automated CT biomarkers for opportunistic prediction of future cardiovascular events and mortality in an asymptomatic screening population: a retrospective cohort study Lancet Digital Health 2020;published online March 2. Available from: https://doi.org/10.1016/S2589-7500(20)30025-X.
[55] Weiss J, et al. Artificial intelligence-derived imaging biomarkers to improve population health. Lancet Digital Health 2020;. Available from: https://doi.org/10.1016/S2589-7500(20)30061-3.
[56] Mori Y, et al. Real-time use of artificial intelligence in identification of diminutive polyps during colonoscopy. Ann Intern Med 2018;169:357–66.
[57] Wang P. et al. Development and validation of a deep learning algorithm for the detection of polyps during colonoscopy. Nat Biomed Eng 2:741–748.
[58] Goshen R, et al. Computer-assisted flagging of individuals at high risk of colorectal cancer in a large health maintenance organization using the Colon-Flag Test. JCO Clin Cancer Inf 2018;2:1–8.
[59] Talboom JS, et al. Big data collision: the internet of things, wearable devices and genomic in the study of neurological traits and disease. Hum Mol Genet 2018;27(R1): R35–9.
[60] Byrom B, et al. Brain monitoring devices in neuroscience clinical research: the potential of remote monitoring using sensors, wearables and mobile devices. Clin Pharmacol Therapeutics 2018;104:59–71.
[61] Ferroni P, et al. Validation of a machine learning approach for venous thromboembolism risk prediction in oncology. Dis Markers 2017;ID:8781379. Available from: https://doi.org/10.1155/2017/8781379.

[62] Schiff GD, et al. Screening for medication errors using an outlier detection system. J Am Med Inf Assoc 2017;24(2):281–7. Available from: https://doi.org/10.1093/jamia/ocw171.
[63] Mackey TK, et al. Twitter-based detection of illegal online sale of prescription opioid. Am J Public Health 2017;107(12):1910–15. Available from: https://doi.org/10.2105/AJPH.2017.303994.

CHAPTER

10

Applications of AI in the psychological and psychiatric fields: what the experts offer us

Summary

Psychiatry is a very significant medical discipline in which human and relational data continue to be predominant over technological data. Even psychology and psychiatry, the "digital," however, has an increasingly important role, and in some respects, certainly positive. Even psychologists and psychiatrists will have to deal more and more often with digital, perhaps to correct or integrate it.

Digital medicine provides us with tools that could help respond to the numerous and complex needs of patients with mental disorders.

Different AI devices, smartphones, and wearable sensors are used to monitor different parameters, with the aim of creating a sort of continuous window on the emotional and behavioral state of patients in the real world, not in the limited one in time and space (artificial) of clinics and hospitals.

The tools that the research has made available are in fact able to analyze the following:

- Autonomic nervous system modulation markers (heart rate variability and breathing patterns, galvanic response, conductance, and skin temperature)
- Vocal analysis of spoken language, in particular, the tone of voice
- Sleep evaluation for duration and quality
- Recognition of facial expressions (especially for the disabled and children)
- Movement, physical activity, communication patterns
- Behavior on social media (interruption, calls, friend requests, choices of image filters)

DOI: https://doi.org/10.1016/B978-0-443-14054-9.00007-7

As part of the sleep study, there are studies comparing wearable devices and "gold standard" monitoring, represented by polysomnography. The devices tested obtained reliable results on total sleep time and lower quality [1].

Many case-control studies, moreover, on small numbers and for short observation periods, have evaluated the motor activity, with ankle oscillometers, of depressed patients. In summary, it turned out that, very predictably, the reduced mood is associated with less physical activity, both in major depression and in bipolar disorders, as well as in schizophrenia.

Speech analysis tools, which are being developed more and more, would be able to diagnose the severity of major depression in the same way as traditional methods such as the Hamilton scale [2].

Through the analysis of texts published on Facebook, it would be possible to predict the onset of the first depressive episode 6 months earlier than the first clinical survey [3].

Similar results would have been obtained from the analysis of photographs posted on Instagram [4].

One of the advantages of digital management of mental disorders would be to cancel the detachment with people who live in rural areas, disadvantaged and low-income countries, and subjects generally "not reachable" by the common health services.

Electronic treatments are, however, not new in the field of mental disorders. There is a vast literature on Internet-based *cognitive-behavioral* therapy for depression, anxiety, and eating disorders, although, in most cases, these programs are based on sessions conducted on a traditional computer screen, not on DI or apps [5].

According to advocates of advanced technologies, digital sensors could, in the future, integrate approaches based on *online counseling* activities, virtual avatars, and cognitive-behavioral therapies, highlighting early signs of relapse, adherence to therapy, and effectiveness of treatment, to allow for a possible pharmacological adjustment or grasp the need for interaction with psychiatric services.

There are apps that offer chatbots that can perform cognitive-behavioral therapies by relating to depressed patients. In studies whose control was a simple informational application on depression symptoms, automated chats have been shown to significantly reduce symptoms of depression and anxiety based on the common PHQ-9 and GAD-7 scales [6].

Other studies have evaluated the bidirectional relationship between neuropsychiatric and cardiovascular diseases, for example, by analyzing heart rate variability, an indicator of autonomic dysfunction. Stress has been found to increase heart rate and reduce its variability. Autonomic dysfunction is also associated with the severity of symptoms in bipolar disorder, acute-phase schizophrenia, and posttraumatic stress disorder [5].

The "digital" in the psychological-psychiatric field: history helps us to understand and recognize potentials and dangers

To realistically evaluate the possible consequences of the introduction of AI in the psychotherapeutic field, we propose a brief historical excursus, illustrating some significant stages without the pretense of completeness and exhaustiveness but with the sole purpose of underlining the worrying dissociation between some applications of artificial intelligence and diagnostic, therapeutic current practice [7].

In the beginning, it was almost a game

The first shy experiment with unexpected results dates back to 1966; Joseph Weizenbaum created a computer program that he called Eliza, which analyzed the statements or questions that the interlocutor typed on the keyboard and answered using generic phrases inspired by common sense and an attitude of human understanding (the researcher's reference was the Rogersian-type psychotherapeutic technique).

For example, if the person typed "*... today I'm sick...*" on the keyboard, the program could answer, "*nowadays it happens to be sick: please tell me*" [8].

The absolutely surprising fact, which after years of studies led Weizenbaum to talk about "Computer Power," was that those who used the Eliza program tended to use it again in moments of tension and declared to have benefited from it, recognizing unusual and unexpected therapeutic abilities on the computer.

The "strange powers" of computers equipped with suitable software were denied, diminished, or ignored by a large part of the psychotherapeutic schools but not by the researchers who continued in their promising, at that time, singular line of research.

Decades later it, went even better with a video game

Interesting and disturbing applications concern the treatment of depression in adolescents; the SPARX (Smart Positive Active Realistic X-factor) software, which looks like a video game, has achieved an improvement in hundreds of adolescents suffering from depression, similar to that shown by adolescents treated directly by the psychotherapist. The sobering fact is that the improvement in mood persisted after three months and even seemed to be superior to traditional therapy in the subgroup with more pronounced symptoms, perhaps because it is easier for young patients to join computerized therapy rather than go to the psychotherapist [9].

Even the very authoritative nice has expressed itself in favor

The quantity and quality of data in favor of computerized Cognitive Behavior Therapy became such that in 2009 the English NICE (National Institute for Clinical Excellence) officially approved the use of this type of therapy in anxiety and depression [10]. But it was just the beginning: cognitive-behavioral therapy administered via the internet has been shown to be effective in various mental disorders, in particular, in depressive disorders, generalized anxiety disorders, panic disorders, obsessive-compulsive disorders, posttraumatic stress disorders, adaptation disorders, chronic pain, phobias, and recently even in tics. [11,12]

Is virtual reality better than psychotherapy?

The need to treat the numerous American soldiers returning from war campaigns in the Middle East and suffering from Posttraumatic Stress Disorder (in some departments, one among six military soldiers) has led to the development of various "IT solutions" in which the soldier distinguished himself for virtual reality software, developed by the University of California in collaboration with the US Army, which led the military soldiers by this important disorder to relive and rework in a protected environment, with adequate responses, to the serious emotional traumas suffered during the war.

Avatars beyond psychotherapy?

Studies on Avatars have broken a historical barrier; that of the therapy of auditory hallucinations, which, as is known, are particularly difficult to treat even with generous use of psychotropic drugs. An important study has shown that an appropriate and programmed Avatar can reduce auditory hallucinations, maintaining a therapeutic effect even after weeks; obviously, according to the authors, the method must be perfected and controlled, but another historical barrier has been overcome by the techniques of computer technology [13].

But the apps are already spreading

The scientific societies and professional organizations of psychiatrists and psychotherapists have prudently not hindered the spread of these new tools, limiting themselves to reaffirming the fundamental importance of the psychotherapeutic relationship. The Apps that diagnose the user's psychic disorder and suggest or direct that some therapies are, however, more and more numerous [14,15]. The data seems to excite a large part of AI advocates, but closer scrutiny should suggest caution or even concern.

For example, a research of a few years ago, funded by various Canadian public institutions, identified 117 "apps" that claimed to be able to help depressed people; however, among these, only 12, equal to 10.2%, seemed to guarantee or exceed minimum quality standards [16].

Given that there are no controls on the quality of the "apps" and there are no mandatory minimum standards to be respected, there is no reason to assume substantial improvements in this very delicate area.

Reflections

The remarkable IT innovations of recent years have radically transformed our world; the manifestations of this change can, without emphasis, be defined as revolutionary; a historical reference repeatedly recalled by various experts is that of the Industrial Revolution in Europe of the 17th and 19th centuries.

In fact, there is no lack of similarities; then and now, albeit on a different scale, various technological innovations complementary to each other made it possible to replace the work of thousands of people; then and now, the technique had been created by a small number of brilliant scientists but was owned by an elite who dictated the rules of the game; then and now the technique was not superior to the intelligent mind and hand skill of man but produced a large number of products of acceptable quality at low cost; then the artisans, and now the intellectuals deluded themselves to be irreplaceable; they became mere workers and now what will happen to health workers and in particular to therapists?

And above all, if digital psychotherapy devices can continue to fly free on the web, how could we protect millions of fragile people susceptible to all kinds of web-guided manipulation?

Perhaps we could save ourselves by creating an international network of validated knowledge and skills capable of controlling the use of these new technologies or a "democratic artificial intelligence" [17].

We hope so, for us, but above all for those who come after us... remembering the Latin phrase *Dum Romae consulitur, Saguntum expugnatur,* which literally translated means "while Rome is discussing, Sagunto is conquered."

Psychological and psychiatric diagnostics through the study of facial expressions

The first significant research on psychological and psychiatric diagnosis through the study of facial expressions dates back to 1969: Paul Ekman, based on some studies carried out on subjects of different ethnic groups, argued that the mimicry of six emotions, namely joy, sadness, anger, fear, surprise, and disgust was uniform throughout the human species and that it was, therefore, possible to study the mimicry of individuals of the most diverse races, identifying their main emotions and subsequently tracing them back to their inner world, and therefore, to their character traits, and their psychological characteristics [18,19].

The careful observation of patients during the interview had been the patrimony of European psychiatrists for over a century but was described masterfully as early as 1872 by the great Charles Darwin in the book *"The Expression of the Emotions in Man and Animals"* [20]. At the end of the 1960s, thanks to Ekman and colleagues, a qualitative leap was attempted, passing from the subjective and nonreproducible interpretation of anthropologists and psychiatrists to a more precise investigation, reproducible and transmissible to other people, such as that allowed by photos and video clips.

From those years an important line of research devoted itself to tracing emotions back to normal and pathological personality traits, and therefore, to providing, to those who knew those interpretative criteria, important tools for formulating hypotheses of psychological, character, but also psychiatric classification.

From the world of research, however, in addition to interesting contributions open to criticism from the entire scientific community, various software for classifying individuals were also prepared, which in the United States were sold to large private companies (including IBM, Amazon, and Microsoft) and even supported by some judges of the US Supreme Court who considered it possible to trace the emotions, and therefore, the personality of the accused [21].

Over the years, also due to evident inconsistencies that emerged with the spread of these methods, an ever-increasing number of scholars questioned Ekman's research. Chen and colleagues, in 2018, showed that Ekman's conclusions were incorrect; the facial expressions of some emotions vary in different races: for example, while expressions of pain are similar between Westerners and Orientals, expressions of pleasure can vary considerably, and the software in use is influenced by the ethnicities of the subjects with whom they have been trained [22].

Furthermore, accurate studies of the physiology of the facial muscles have shown how the 43 muscles present can contract or relax, generating dozens of expressions that are difficult to classify and do not coincide with the six described Ekmann emotions. A massive review of the literature on the subject was recently published; about 1000 articles were evaluated and compared by a group of psychologist experts.

The conclusions are clear and unambiguous; **there is no certain proof that the emotional state of a person can be deciphered from simple facial expressions** [23].

Some large companies specializing in character "profiling" (notably Affectiva of Boston [21]), however, argue that their software provides correct results in the vast majority of cases, and this is what the clients need.

In other words, in the world of business, industry, and commerce, scientifically questionable applications are emerging, imperfect, and often burdened with important biases, but very useful for selecting people with the characteristics desired by those who pay.

That's the way the world goes, or, perhaps better, "So it is, if you like" [24].

References

[1] Mantua J, et al. Reliability of sleep measures from four personal health monitoring devices compared to research-based actigraphy and polysomnography. Sensors (Basel), 16. 2016. p. 646.

[2] Cannizzaro M, et al. Voice acoustical measurements of the severity of major depression. Brain Cogn 2004;56:30–5.

[3] Eichstaedt JC, Smith RJ, Merchant RM, et al. Facebook language predicts depression in medical records. Proc Natl Acad Sci USA 2018;115:11203–8.

[4] Reece AG, Danforth CM. Instagram photos reveal predictive markers of depression. EPJ Data Sci 2017;6:15.

[5] Reinertsen E, Clifford GD. A review of physiological and behavioral monitoring with digital sensors for neuropsychiatric illnesses. Physiol Meas 2018;39 05TR01 (38 pp).

[6] Fitzpatrick KK, Darcy A, Vierhile M. Delivering cognitive behavior therapy to young adults with symptoms of depression and anxiety using a fully automated conversational agent (Woebot): a randomized controlled trial. JMIR Ment Health 2017;4:e19.

[7] Maddox TM, Rumsfeld JS, Payne PR. Questions for artificial intelligence in health care JAMA. Published online Dec 2018;10. Available from: https://doi.org/10.1001/jama.2018.189323.

[8] Weizenbaum J. ELIZA – a computer program for the study of natural language communication between man and machine Commun Assoc Comput Mach 1966;9:36–457. Available from: https://www.csee.umbc.edu/courses/331/papers/eliza.html.

[9] Sally NM, Karolina S, et al. The effectiveness of SPARX, a computerized self help intervention for adolescents seeking help for depression: randomized controlled non-inferiority trial. BMJ 2012;344:e2598.

[10] National Institute for Clinical Excellence. Computerized cognitive behavior therapy for depression and anxiety. Technol Appraisal 2009;97. Available from: http://www.nice.org.uk/nicemedia/pdf/TA097guidance.pdf.

[11] Kumar V, Satta Y, et al. The effectiveness of internet-based cognitive behavioral therapy in treatment of psychiatric disorders. Cureus 2017;9(8):e1626. Available from: https://doi.org/10.7759/cureus.1626. Open Access Review Article.

[12] Daly T. Mobile apps help me manage my tics. BMJ 2019;366:l2415. Available from: https://doi.org/10.1136/bmj.l2415.

[13] Craig Tom KJ, Rus-Calafell M, et al. AVATAR therapy for auditory verbal hallucinations in people with psychosis: a single-blind, randomised controlled trial. Lancet Psychiatry. 2018;5(1):31–40. Available from: https://doi.org/10.1016/S2215-0366(17)30427-3.

[14] Leigh S, Flatt S. App-based psychological interventions: friend or foe? Evid Based Ment Health 2015;18(4):97–9.

[15] Burns MN, Begale M, Duffecy J, Gergle D, Karr CJ, Giangrande E, et al. Harnessing context sensing to develop a mobile intervention for depression. J Med Internet Res 2011;13(3):e55. https://www.jmir.org/2011/3/e55/.
[16] Huguet A, Rao S, et al. A systematic review of cognitive behavioral therapy and behavioral activation apps for depression. PLoS One 2016;11(5):e0154248. Available from: https://doi.org/10.1371/journal.pone.0154248.
[17] Santoro Eugenio. Artificial intelligence, medicine, society. Res Pract 2018;34 27-27.
[18] Ekman P, Sorenson E, et al. Pan-cultural elements in facial displays of emotion. Science 1969;164(3875):86–8 Apr 4.
[19] Ekman P, Friesen W:. Constants across cultures in the face and emotion. J Pers Soc Psychol 1971;17(2):124–9 Feb.
[20] Darwin C. The expression of emotions in humans and animals Bollati Boringhieri. Ed. Turin; 2012.
[21] Heaven D. Expression of doubt. Nature 2020;58:502–4.
[22] Chen C, Crivelli C, Jack R, et al. Distinct facial expressions represent pain and pleasure across cultures. Proc Natl Acad Sci USA 2018;115(43):E10013–21.
[23] Barrett LF, Adolphs R. Emotional expressions reconsidered: challenges to inferring emotion from human facial movements. Psychol Sci Publ Interest 2019;20(1):1–68.
[24] Pirandello L. So it is (if you like). Milan: Garzanti Ed.; 2014.

CHAPTER

11

From personalized medicine to precision medicine

Summary

Personalized medicine was born with Hippocrates and was dormant in the Middle Ages; it experienced a new spring with the 19th clinic based on biological research, and it was eclipsed by the overwhelming, pervasive affirmation of the technique.

The bursting development of genetics has given us new important tools to return to the person: to us the responsibility of using them also as a tool for enhancing the "humanistic" component of care.

Doctors have always practiced personalized medicine, intended as the use of professional experience to adapt the best literature data to the individual patient. At the end of the 19th century, Canadian doctor William Osler affirmed the concept with the famous motto: "It is much more important to know which type of patient has a disease than which disease a patient has." Moreover, medical science has had a predominantly population approach, favoring, especially in the pharmacological field, the response of the "average patient" over the particular case. Overall, the results were of great value, although millions of people take medications every day that have little or no efficacy in their case while at risk of adverse events. According to nature, the ten drugs with the highest turnover in the United States work, at best, in one out of four patients, at worst in one in fifty, so much so that we talk about "imprecision medicine (PM)" [1].

PM aims to analyze all the dimensions of the single individual to "tailor" the treatment based on individual characteristics. Indeed, it can take advantage of the enormous developments in molecular biology and the creation of enormous databases whose information can be processed by AI systems to objectify and quantify the heterogeneous nature of most diseases and the phenotypic variability of individuals at the genomic level, epigenomics, transcriptomics, proteomics, and metabolomics, the so-called "panomics."

DOI: https://doi.org/10.1016/B978-0-443-14054-9.00004-1

The term PM was initially coined in oncology, following the fact that neoplasms may depend in their development on genomic alterations, which can become the target of specific inhibitory treatments capable of acting with a high probability of success against the neoplasm and causing limited harm to the patient [2]. The identification of more effective therapies, as *targeted* toward patients with responsive subtypes, avoids exposing non-responsive subjects to useless side effects, with important prognostic, safety, and cost implications. There are currently many drugs being tested and many already on the market used with this approach. Just think of imatinib for the treatment of chronic myeloid leukemia in the presence of BCR-ABL mutation or of trastuzumab in HER2-positive breast cancers.

The advances in PM have also extended to other areas; for example, in cardiology, they have made it possible to discover 46 different variables of heart failure with preserved systolic function, a notoriously inhomogeneous population, identifying three specific groups that have shown extremely different clinical outcomes [3].

Limitations of precision medicine

Especially in the oncology field, the availability of molecular tests for a large number of biomarkers has produced potentially useful new drugs, but also a greater number of subtypes of neoplastic diseases to be treated. In practice, as stated by the epidemiologists F. Trotta and G. Traversa, through progressive segmentation, even high-prevalence tumors are treated today as rare diseases. *This makes us reflect on the very usefulness of PM, and in any case, on the need to place a limit on the application of molecular biology to clinical practice.* In fact, the authors wonder to what extent it is worthwhile to characterize the neoplastic disease so as to make it so rare as to jeopardize the feasibility of clinical studies to validate a treatment. Furthermore, often the biomarkers involved in the mechanism of action of a single drug are numerous, and therefore, more than one is predictive of clinical efficacy; in practice, the drug becomes not *targeted* but *multitargeted*. The heterogeneity of the loci also implies an expansion of the decision-making algorithms for the prediction of abnormal pathophysiological pathways in translation into diseases. Many limits of PM are well identified by Perrone [4] and summarized next:

- What meaning should be assigned to the information that genetic sequencing produces, taking into account the mutability of genetic expression and the dynamism of cancer cells, capable of transforming and defending themselves by becoming resistant; what meaning to attribute to the "background noise" to the possible artifacts?

- What reliability and impact do the different reading systems of massive sequencing tests have?
- On what evidence to arrive at the registration of new drugs?
- What sustainability for equal accessibility to therapeutic innovations?
- What information to patients? How to avoid unjustified expectations, which can disappoint patients and family members?

Finally, the high price of anticancer drugs, in the absence of a strategy aimed at their lowering, will be difficult to sustain. The fear of many experts is that funding for PM (the Obama administration has allocated $ 215 million for 2016 to the *Precision Medicine Initiative*) could penalize public health strategies, individualizing what should be a social challenge [5].

The need for new methodologies

PM requires new methodologies to evaluate effectiveness, outcomes, safety, sustainability, and as regards the cultural impact in terms of training/information for health personnel and citizens [6]. Classical studies, fundamental to highlighting the efficacy of drugs at the population level, do not reconcile with PM. Different methodologies are needed, such as the *umbrella trial* and the *basket trial*.

In the first, patients with a certain type of neoplasm are studied to evaluate the presence of a series of biomarkers, and on this basis, distributed in the treatment arms with the corresponding drugs, that is, each drug is associated with the specific biomarker.

In the second methodology, patients are recruited only on the basis of molecular characteristics, and therefore, tumors also originating in different organs are allocated in the same treatment arms. To overcome the limitations of numbers due to the low prevalence of many molecular markers, the so-called *master* protocols are also used, which involve the use of diagnostic test batteries to assign patients to a network of ongoing trials.

Moreover, these new study designs are believed to be especially useful in the early phase of drug development. For the demonstration of clinical efficacy, the traditional clinical *trial* design remains the reference standard [7,8].

Final considerations

PM represents a novelty capable of determining great changes at the experimental and clinical level, for example, to reduce waste and iatrogenic risks. Furthermore, there are many questions that will have to be answered, as regards the clinical implications ("from the genetic laboratory to the patient's bed"), but also legal, ethical, economic, and political. Above all, General Medicine, an ontologically oriented discipline to complexity, to "true" holism, characterized by a basically de-medicalizing approach, is able to highlight the limits of PM.

It should be emphasized that the new molecular technologies associated with AI can allow a redefinition of medical nosography [9] *and the construction of new diseases, or in any case. innumerable subtypes of the same* (from *targeted therapy* to *targeted research of diseases*). Machine learning has made it possible, for example, to identify new allergy phenotypes in children, and through different IgE patterns, to predict the risk of asthma with greater precision [10].

This will allow the use of targeted and highly effective therapies, but in some cases, the price to pay will be an expansion of medicalization, a sort of "holistic medicalization." Citizens must therefore be informed about the limits and potential of the different scientific approaches, sometimes defended by a "precision prevention," based on molecular fingerprints that can degenerate into a prediction as an end in itself, without practical implications but with enormous psychological effects and social. The term genocondriac has already been coined! [11].

Another fundamental aspect is that PM is essentially based on a techno-scientific perspective. Its other possible applications, such as the technology of electronic health records, medical devices and wearable sensors, and information deriving from *social media* and other sources [12] are able to provide an important amount of data to create a sort of "medical avatar" of each citizen, but daily experience teaches that assistance" precisely "tailored to the individual must also include (above all?) the patient's subjective life circumstances, his *ability to* cope, his values, his personality, his fears and hopes, dimensions that are difficult to transfer to a *database*.

There are undoubtedly interesting pilot projects, which, however, propose patient-oriented approaches, for example, using genetic and physiological big data to propose lifestyle modifications based on the risk of diabetes or hypertension or specific therapeutic variations based on information from individual pharmacogenetics [13].

To share an ever greater attention to the understanding of the individual, who must also ensure a global consideration beyond basic biology, we agree with R. Ziegelstein, who proposed to add the suffix -omics to the word person, coining the term *personomics* [14]. The claim to create an ever more exact science of people's lives to objectify and "calculate" it, in an increasingly reductionist perspective, risks going beyond the desirable personalization of treatments toward a complex phenotypic determinism but at the same time "trivializing," fatalistic, and ultimately demotivating for the single individual.

References

[1] Schork NJ. Personalized medicine. Time for one-person trials. Nature 2015;520:609–11.

[2] De Braud F. Precision, hopes, risks and possible objectives (interview). <http://www.forward.recentiprogressi.it>.

[3] Pitt GS. Cardiovascular precision medicine: hope or hype? Eur Heart J 2015;36:1842–3.

[4] Perrone F. The unknowns of a journey into knowledge. <http://www.forward.recenti.progressi.it>.

[5] Bayer R, Galea S. Public health in the precision medicine era. N Engl J Med 2015;373:499–501.

[6] Hunter DJ. Uncertainty in the era of precision medicine. N Engl J Med 2016;375:711–13.

[7] Trotta F, Traversa G. The search for precision among umbrellas and baskets. <http://www.forward.recentiprogressi.it>.

[8] Biankin AV, Piantadosi S, Hollingsworth SJ. Patient-centered trials for the therapeutic development in precision oncology. Nature 2015;526:361–70.

[9] Miernezami R. Preparing for precision medicine. N Engl J Med 2013;366(6):489–91.

[10] Fontanella S, et al. Machine learning to identify pairwise interactions between specific IgE antibodies and their association with asthma: a cross-sectional analysis within a population-based birth cohort. PLoS Med 2018;15(11):e1002691. Available from: https://doi.org/10.1371/journal.pmed.1002691.

[11] Collecchia G. From personalized medicine to precision medicine. IsF 2017;41 (n. 2):19–22.

[12] Parikh R, et al. Beyond genes and molecules - a precision delivery initiative for precision medicine. N Engl J Med 2017;376:1609–12.

[13] Mahoney MR, Asch SM. Humanwide: a comprehensive data base for precision health in primary care. Ann Fam Med 2019;17:273. Available from: https://doi.org/10.1370/afm.2342.

[14] Ziegelstein RC. Personomics. JAMA Intern Med 2015;175:888–9.

CHAPTER

12

The search for new drugs

Summary

Many Artificial Intelligence (AI) techniques are used for drug innovation. From the sophisticated research of biomedical literature for the analysis of data relating to millions of molecular structures to the interpretation of biological data of diseases, for which to propose new active ingredients, to the design and implementation of possible new molecules, from screening vast libraries to predicting structures via AlphaFold[1] to the identification of new indications for already existing drugs, to the prediction of a possible favorable response to new active ingredients.

The discovery by the robot Eve, a project of the universities of Cambridge and Manchester, of an antimalarial product in the composition of a toothpaste, has emphasized the interest in the use of AI in the discovery of new drugs, with a long list of start-ups and partnerships with the drug industry. There are dozens of multinational drug companies that have established relationships with both the IT giants (Google with DeepMind and IBM with Watson for Drug Discovery) and with over a hundred start-ups, with an ever-expanding increase.

Using in silico simulations, based on the chemical structure of the molecules, it is also possible to predict possible side effects and toxicity, evaluate the right dose for experimental drugs and discover new interactions between drugs already on the market [1].

In particular, there is hope that the prediction of toxicity through machine learning algorithms will allow reducing preclinical tests on

1 AlphaFold represented a breakthrough in the key task of protein folding, which involves predicting the 3D structure of a protein from its chemical sequence. Improvements in protein structure prediction can provide mechanistic insight into a range of phenomena, such as drug–protein interactions or the effects of mutations.

DOI: https://doi.org/10.1016/B978-0-443-14054-9.00012-0

animals [2] and the times of experiments[2], for example, through the "organ on chips" technology, which uses stem cells to reconstruct mini-organs in the laboratory. Moreover, we are still far from the realization of a "body on chip," integration of the various organs to reconstruct the complexity of the organism [3].

Other applications are the identification of possible candidates for clinical trials on the basis of clinical or genetic data (see paragraph 11 on precision medicine) and the target of patients who could benefit from the new treatments [4].

A particularly important area is that of antibiotics, given the increasing, rapid, and widespread appearance of resistant bacteria, which according to many experts, will be one of the main causes of death within a few decades. In fact, it is estimated that, in the absence of powerful and targeted action, there will be 10 million deaths a year worldwide between now and 2050 [5]. In this regard, some scientists from the Massachusetts Institute of Technology (MIT) in Boston have used a model of a deep neural network (deep learning) trained to predict, for a number equal to 108, the molecular structure of possible active ingredients able to interact with pathogens. The result was halicin, a compound substantially different from conventional antibiotics from a structural point of view and which has shown an excellent bactericidal capacity on a broad spectrum of pathogenic strains, such as *Mycobacterium tuberculosis*, various species of *carbapenem-resistant enterobacteria*, Clostridioides Difficile *and Acinetobacter baumannii* [6]. The researchers also highlighted eight other molecules structurally distant from common antibiotics. AI, therefore proposes itself as an effective tool for the production of weapons in the fight against antibiotic resistance.

References

[1] Topol E. High-performances medicine: the convergence of human and artificial intelligence. Nat Med 2019;25:44–56.
[2] Luechtefeld T, et al. Machine learning of toxicological big data enables read-across structure activity relationship (RASAR) outperforming animal test reproducibility. Toxicology 2018;165:198–212.
[3] Cerati F. Digital guinea pigs for kind searches. NOVA Il Sole 24 ORE, February 4, 2018.
[4] De Biase L. Artificial intelligence in medicine. Digital Health: from doing to treating. http://www.cdti.org
[5] O'Neill J. Antimicrobial resistance: tackling a crisis for the Health and Wealth of Nations (Review on Antimicrobial Resistance, 2014).
[6] Stokes JM, et al. A deep learning approach to antibiotic discovery. Cell 2020;180:688–702.

2 In recent years, the average price to market a new drug is close to 2.5 billion dollars, with a delivery date of around 10–15 years.

CHAPTER

13

Digital therapeutics

Summary

The diagnostic and therapeutic approach of digital devices, in particular in the field of cognitive-behavioral therapy (CBT)[1], has paved the way for the development of digital therapies, abbreviated as (DTx). These can be defined as therapeutic tools, based on evidence of efficacy, with clinical objectives: they consist of apps, wearable devices, and software, which alone or in association with other interventions, have the purpose of preventing, managing, or treating chronic or related diseases to behavioral or psychological factors (see Table 13.1).

In Italy, no DTx is currently available for clinical use. What is considered as such are, in reality, the *Patient Support Programs (PSP)*, interventions of various types, which do not have intrinsic therapeutic properties, but rather the purpose of optimizing the treatment that the patient is carrying out. In particular, their efficacy is rarely subjected to experimental verification through randomized and controlled clinical trials.

DTx is sometimes assimilated to the so-called "digital medicines" that is, drugs with an integrated sensor that can be assimilated by the human body, which after ingestion, is activated in the stomach upon contact with gastric juices. The activation determines the release of an electrical signal, which is transmitted to a special patch, applied on the patient's body, is further transmitted via Bluetooth to the app of the patient's smartphone, which signals to the doctor or caregiver the actual intake of the drug, as in the case of *Abilify MyCite*, an aripiprazole tablet used in the treatment of schizophrenia and other psychiatric conditions, and approved in 2017 by the FDA. They are, therefore, tools for monitoring adherence, not real therapy.

1 form of psychotherapy based on the assumption that there is a close relationship between thoughts, emotions, and behaviors, and that by operating on thoughts it is possible to change behaviors

DOI: https://doi.org/10.1016/B978-0-443-14054-9.00018-1

TABLE 13.1 Selection of therapeutic contexts in which DTx has the greatest development.

- Diabetes mellitus
- Hypertension
- Asthma and COPD
- Obesity
- Insomnia
- Depression
- Addictions (smoking, alcohol, drugs)
- Adverse reactions from antineoplastic drugs
- ADHD - Attention Deficit Hyperactivity Disorders
- Autism Spectrum Disorders
- Cognitive impairment

Mode of action

As described by G. Recchia [1], what differentiates DTx from the drug is the active principle, that is the element of the medication responsible for the clinical effect. In the case of a drug, a chemical or a protein molecule, and in the case of DTx, a software or an algorithm, which, unlike traditional drugs, in order to involve the patient, in most cases, takes on a playful form, such as apps, video games, and sensors. In fact, DTxs represent a direct interface with the patient, and their role is potentially fundamental in all phases of product development, from design to evaluation of clinical outcomes, in particular, safety.

Continuing in the analogy with drugs, the "excipients" give shape to the active ingredient and promote its intake, making it as digitally bioavailable as possible.

They include, for example, apps for patient rewarding, reminders for taking DTx, and complementary therapies, modules for connecting the patient with their doctor and with other patients with the same indication.

Therapeutic indications

DTxs can be used to treat a variety of diseases, either independently or in combination with drugs, devices, or other therapies.

An example of the first category are apps capable of reducing glycated hemoglobin levels in diabetic patients, promoting proper nutrition and more constant physical activity. The best known is probably Bluestar Diabetes, one of the first DTxs developed.

The so-called "intelligent inhalers" (such as Propeller) belong to the second category, which, thanks to the help of sensors connected to software, are able to improve adherence to inhalation therapy [2].

The area in which DTxs have mostly developed, especially in the United States, is that of mental health. Of the more than 700 different startup proposals, the greatest ability to attract investments is concentrated in the area of DTxs (30%) [3].

The so-called "intelligent pills" are at an advanced stage, which, equipped with specific sensors, monitor the patient from inside the body and autonomously decide the release of drugs based on the data collected. The pill, 3D printed and covered with a material that protects it from gastric acidity, is able to remain in the stomach for at least a month and can gradually release the drug it contains. The sensors it is equipped with allow, on the one hand, to monitor the gastric environment, thus allowing identification of the first signs of disease, and on the other, to increase adherence to pharmacological treatments to which a patient is subjected. Through the Bluetooth connection it is also possible to use an app installed on the smartphone to modulate the amount of drug that the smart pill must release. Other advantages that researchers expect are the possibility of administering drugs that otherwise would have to be injected and that of promptly treating patients who have been diagnosed with a certain pathology. The system may be useful for promptly treating patients suffering from particular pathologies, for example, immunosuppressed, in which the pill could detect the presence of infection early and release the necessary antibiotic. The smart pill is currently being tested in pigs, but within the next two years, it will be tested on humans [4].

Regulation

Like pharmacological drugs, DTx, before being included in clinical practice, are subject to regulation by the competent authorities (Food and drug administration (FDA) in the United States and European Medicines Agency (EMA) in Europe, in order to measure its clinical efficacy and safety, through randomized and controlled trials, on measurable clinical outcomes. At the end of the study process the DTx can be classified as a medical device. Directives will, however, soon become more stringent, and DTxs will have to comply with stricter regulatory requirements. In particular, they must be equipped with a quality management system, postmarketing analysis, a summary of safety, and clinical performance [5].

At the moment, no prescription DTxs are authorized and prescribed by the doctor and used by the patient in Italy, where specific legislation on digital health is lacking. In Italy, it is not yet clear whether the Ministry of Health (as medical devices) or the Italian Medicines Agency (AIFA) as therapy will be responsible for DTx.

In the other European countries, limited experiences are present in Germany, the United Kingdom and France.

Pending specific guidelines, developers can address scientific and approval questions to the EMA, which has recently proposed a document, subject to public consultation, which includes digital areas of interest to create an adequate regulatory system in the pharmaceutical field [6].

In the United Kingdom, the NICE (National Institute for Health and Care Excellence), in agreement with the NHS England and the NHS digital, has published *guidance* for manufacturers necessary to accredit their products and make them prescribable and reimbursable, taking into account the effectiveness and economic impact [7].

Recently NICE published a draft in which it states that DTxs, in the form of apps, wearable devices, and online programs, can be used as complementary tools to help people change behaviors, in particular, to become more physically active, manage weight, quit smoking, reduce alcohol intake, or avoid unsafe sexual behavior. Experts invite doctors to inform them about possible complications and point out the lack of reliable data on the effectiveness of TD when used in the absence of professional advice. In practice, a personalized assessment is recommended, case by case [8].

The implementation in Germany of the new Digital Health Service Act/Digitale-Versrgung-Gesetz-DVG, which allows doctors to prescribe DTxs that have passed the safety and scientific checks by the regulatory body (BfArM)[2] and allows reimbursement to citizens covered by public health insurance, represents a fundamental step forward for the entry of DTxs into medical practice [9,10].

In the United States, the FDA-approved ReSET app offers a CBT for treating opioid addiction and abuse problems. In the pivotal trial, DTx demonstrated, versus CBT performed by healthcare professionals, a significant improvement (40.3% vs 17.6%) in the abstinence rate from alcohol, cocaine, and marijuana, and adherence to treatment in the extra regime hospital [11].

Conclusions

Currently, the DTxs still have many open questions and limited certainties. In particular, the level of studies needed to evaluate their effectiveness is not clear: controlled and randomized clinical trials or real-world evidence? [12] In fact, many wonder if the classic studies are able to cope with the dynamic, ever-accelerating digital technological development [13]. There is also a lack of certainty about the regulation of data protection responsibility.

2 Entrepreneurs of digital therapies will have to apply for their products to be introduced in the register on a preliminary basis for 12 months, during which time they will be subjected to careful testing

As stated by E. Santoro, they suffer in general from the ambiguities that have characterized the field of Digital Health [14]. This has led, especially doctors, not to see their potential, both in clinical and economic terms, although they have already been in use for years and positively evaluated by prestigious agencies such as NICE in England. It is desirable that these innovations, validated for safety, efficacy, and efficiency, can be used and integrated into clinical practice, for better healthcare, in particular for more widespread, equitable, and sustainable access to the health system, through the possible greater involvement of people for an adequate response to the many unmet needs.

References

[1] https://psichiatriadigitale.it/2019/10/03/terapie-digitali-che-cosa-non-sono-che-cosa-sono-che-cosa-saranno/.
[2] Santoro E. A fast track for digital therapy. Forw (R) Evol 2019;4.
[3] https://medium.com/what-if-ventures/approaching-1-000-mental-health-startups-in-2020-d344c822f757.
[4] Santoro E. http://magazine.familyhealth.it/2019/03/07/le-pillole-intelligenti-aiutano-controllo-del-farmaco-delle-malattie/.
[5] Agricola E, Di, Marzo M. Digital therapies: the European and national regulatory landscape. Forw (R) Evol 2019;4.
[6] https://www.ema.europa.eu/en/documents/regulatory-procedural-guideline/ema-regulatory-science-2025-strategic-reflection_en.pdf.
[7] National Institute for Health and Care Excellence. Evidence standards framework for digital health technologies London: National Institute for Health and Care Excellence; 2018. https://www.nice.org.uk/about/what-we-do/our-programmes/evidence-standards-framework-for-digital-health-technologies.
[8] https://www.nice.org.uk/news/article/digital-and-mobile-interventions-could-support-regular-health-services-in-helping-people-stop-smoking-and-reduce-their-risk-of-obesity-says-nice?utm_medium = social&utm_source = twitter&utm_campaign = phgdigitalinterventionsbehaviour.
[9] bundesgesundheitsministerium.de/digital-healthcare-act.html.
[10] https://digitalhealthitalia.com/il-futuro-delle-terapie-digitali-in-europa-parte-dalla-germania/.
[11] Waltz E. Pear approval signals FDA readness for digital treatments. Nat Biotechnol 2018;36(6):481–2.
[12] De Fiore L. Revolution or evolution? Open questions on the technological and digital transformation of health. Forw (R) Evol 2019;4.
[13] Shaywitz D. Will real world performance replace RCTs as healthcare's most important standard? Forbes, May 2018. https://www.forbes.com/sites/davidshaywitz/2018/05/11/will-real-world-performance-replace-rcts-as-healthcares-most-important-standard/.
[14] https://medicoepaziente.it/2020/digital-therapeutics-cosa-sono-e-come-funzionano-le-terapie-digitali/.

CHAPTER

14

Virtual reality: the great therapeutic potential and the possible risks

Summary

Imagination is the first source of human happiness" (Giacomo Leopardi: Zibaldone), and virtual reality is, in fact, today's expression of daydreams of every age and civilization. From a medical point of view, virtual reality uses little known and used potential of our mind to cure some of our states of suffering. The potentials are many and exciting, the risks many and rather serious, both for improper use and because it uses very reactive but little-known properties of the human mind.

The term "virtual reality" was coined a few decades ago when computers began to simulate real environments and events. In recent years, however, the great advances in digital technology have radically changed the characteristics of virtual reality software and devices; in fact, the most up-to-date devices and software allow users to immerse themselves deeply in a parallel pseudo-reality while maintaining partial contact with the real world.

This new type of experience is characterized by sensory information that is interpreted as perceptions coming from the virtual environment, with the possibility of performing various types of actions in the new environment: the subject has the sensation of being really present in that new environment.

Presence, or rather the subjective experience of being in a specific environment, is one of the most characteristic elements of digital reality: the subject feels involved and inserted in the digital environment and experiences a psychic process of dissociation between perceptual data and cognitive processes; in fact he continues to receive environmental stimuli through the various sensory channels (visual, auditory, tactile,

DOI: https://doi.org/10.1016/B978-0-443-14054-9.00001-6

proprioceptive, cenesthetic) but also in the same time interval other stimuli, once only visual and auditory but now also proprioceptive, and cenesthetic from digital reality [1,2].

In nonimmersive virtual reality, the user is instead simply in front of a monitor, which acts as a window on the three-dimensional world with which the user can interact through special joysticks.

What happens when you enter the virtual-digital neo-reality

The entry into virtual reality begins as soon as the subject feels new and original perceptions of space: there is a particular and crucial phase called the "conversion phase" in which the subject feels a different spatial location and "feels" that he has entered a different environment, to be involved or even protagonist of a different reality; it is not a question of illusion, much less hallucination but rather of identification.

The following are noteworthy:

Illusion is a transformation of real perceptions in which the subject, however, maintains a critical sense

Hallucination is the false perception of sensory stimuli in the absence of stimuli or in the presence of insignificant stimuli; the subject, in this case, does not maintain a critical sense

Identification is a process of rational and emotional identification with another subject [3]

In the experiences of digital neo-reality, we can distinguish a space-time identification, a body identification, and an illusion of reality.

The subject maintains cognitive awareness but feels he is different in a different environment: experiments with digital avatars have shown that the subject feels a more or less intense and profound identification with the avatar. This phenomenon can favor important changes that could have therapeutic value. For example, it is possible to act on the neuromotor level to promote rehabilitation in subjects affected by ischemic cerebral hemorrhagic attacks, as well as it is possible to reduce auditory hallucinations thanks to suitably created avatars [4].

The original characteristics of the mental processes in people involved in virtual reality experiences

It is important to describe, with the help of experts, the original characteristics of mental processes in virtual reality experiences because these peculiarities allow us to understand both the great potential and the possible serious risks [1,2].

1. The first important characteristic is that of the dual and simultaneous perceptual input, virtual and real; as underlined, this is not an illusion nor hallucination but a new enriched reality that generates new psychic states.
2. This neo-reality is a small new world in which the individual is able to try new experiences and elaborate new thoughts.
3. The interaction creates subjective phenomena of suggestion and a rational and creative mind, and the prevalence of one or the other accounts for the wide range of effects of virtual reality: in some subjects they can be limited to a fantasy experience, but in other subjects it can generate extended body and mind experiences with a temporary enhancement of the creative faculties.

Potential adverse effects

The above allows us to understand how digital virtual reality can be a therapeutic tool of great potential but with possible side effects that are also very serious.

The critical element of these experiences is the psychic process of dissociation between perceptual data and cognitive processes; this process is the basis of the great potential of this tool, but it is also the most critical element that can lead to serious imbalances in vulnerable subjects who can experience extremely distressing experiences of depersonalization and derealization with varying intensity. Let us remember that the latter are real psychopathological pictures with important somatic manifestations: we briefly recall their characteristics here.

Depersonalization/Derealization disorder is characterized by experiences of unreality, detachment, by feeling like external observers with respect to one's thoughts, feelings, sensations, and to one's body or one's actions; or, alternatively or in association, from experiences of unreality and detachment from the surrounding environment (people or objects are perceived as unreal, dreamlike, foggy, inanimate). These disorders are generally described with strong anxiety and participation and are frequently accompanied by somatic symptoms like tachycardia, polypnea, and sweating.

These disorders are not attributable to mental structures of a psychotic type but can also occur in "normal" people, although generally very anxious: in the presence of structural alterations, even mild ones of a psychotic type (loss of contact with reality, impairment of self-critical abilities), virtual reality experiences could precipitate psychotic imbalances [3].

Didactic applications: the training of doctors

Virtual reality is also used for training doctors, for example, in the study of anatomy, because it provides a perspective beyond the classic two-dimensional (paper) and static (autopsy) approaches. The method, in fact, allows the creation of 3D images of the human body that can be manipulated, rotated, expanded, and observed from multiple positions. In addition, students can examine the functioning systems, with the heart beating, the blood flowing, and the diaphragm rising and falling (). Virtual reality-based learning models have been introduced in several universities, from California to Australia.

In the United Kingdom, the University of Leeds uses patient holograms with which students can interact and communicate through immersive viewers. The same technology can be used for the training of young surgeons and also as an aid for more experienced professionals who are developing new surgical techniques.

Training qualified personnel is particularly useful in less developed countries, where training is more difficult to perform and surgeons are often few in number.

Other uses may be in *palliative care.* At the Royal Trinity Hospice, virtual reality has been used in patients with very severe ailments, such as those with amyotrophic lateral sclerosis, to provide experiences such as swimming with dolphins.

The barriers to practical use

The main obstacles to the expansion of virtual reality are costs, even if the increase in the market could allow a progressive reduction.

Furthermore, medicine, and surgery in particular, are practical professions: it is one thing to simulate a skin incision; it is quite another to perform it directly. The gap could be resolved using haptic technologies, such as gloves that allow you to receive tactile sensations while performing a surgical operation using a robot.

Augmented reality

Virtual reality must be distinguished from augmented reality: in the first, the wearer is transported to a completely different environment, a pseudo-reality, into which the user can "enter" thanks to special visors and gloves equipped with sensors used for movements, or suits.

In the second, the operator remains in the real environment, but his sensory perception is enriched through virtual objects thanks to the use of sensors and algorithms that allow superimposing computer-generated 3D images on the real world, generating information that would not be perceptible with the five senses, for example, for the surgeon, the three-dimensional image of the organ to be operated on.

Augmented reality has also been used above all in the surgical field, to facilitate the planning of operations and follow the progress of operations [5].

CT and MRI are used to create 3D reconstructions of the human anatomy visible by operators through special viewers that allow you to perform simulations in the preparation for the surgery or during the same. Augmented reality has been used mainly in the hepatobiliary [1–5] and facial skull areas [6,7].

The enhancement of our cognitive faculties and the management of our emotions

In psychically balanced individuals, virtual reality experiences can involve a cognitive-emotional-motivational involvement that can significantly broaden one's horizon of knowledge but also of skills and attitudes. From a cognitive point of view, we can speak of "extended minds," and from an emotional point of view, it is possible to make subjects aware of their emotions by also simulating the consequences of various ways of expressing and managing them.

In this context, the potential, also and especially of a pedagogical-formative-training type, is truly remarkable.

The therapeutic applications of virtual-digital reality

We will examine the main areas of therapeutic applications, remembering that to avoid unwanted consequences, it is essential that virtual reality therapeutic devices are used by highly qualified personnel [7].

Neurological diseases

In the neurological field, the attempts for the therapeutic use of virtual reality date back several decades ago; it is necessary to recall at least the brilliant experiences of the neuroscientist Ramachandran, who, with simple cardboard boxes and mirrors managed to obtain good results in the phenomenon of the phantom limb and to understand better the complex neuropsychological mechanisms of anosognosia [8].

Since then, knowledge and techniques have improved considerably: in the current therapeutic use, four phases can be distinguished:

1. Personalized assessment of the patient's deficits and recovery potential
2. Drafting of a personalized exercise program with psychomotor tasks compatible with residual abilities.
3. Beginning of therapy with continuous feedback and passage to gradually more complex and tiring exercises.
4. Mechanisms of gratification and reward (see operant conditioning) that support the patient in the difficult rehabilitation process

The virtual reality devices in Parkinson's are particularly interesting, where the home environment is reconstructed and the patients are trained to improve their perception of space, and thereby, improve their motor coordination and balance mechanisms; in these cases, a continuous feedback with positive reinforcements is very important [2].

Applications in psychiatry

Phobias

The use of virtual reality in the therapy of phobias gives encouraging results; the therapeutic activity is based on the gradual exposure to the negative stimulus, which is attenuated as much as possible.

The visual and auditory channels are mainly used.

The patient's psycho-emotional response tends to diminish over time: in the most favorable cases, a cognitive restructuring occurs, following which the patient attributes a different meaning to the noxious stimulus, toward which the quality of the response also changes accordingly.

Post traumatic stress disorder

Posttraumatic Stress Disorder is characterized by Altered Vigilance with exaggerated alarm responses even for trivial stimuli, Sleep Disorders, Difficulty Concentration, Irritability, and Reckless and sometimes self-destructive behavior; it can arise in many circumstances but is particularly frequent in those who live or work in situations of war or serious social unrest.

The results obtained by virtual reality devices in this disorder are encouraging. The first experiences date back to 1996 with Vietnam veterans and are now further improved in applications for veterans of Afghanistan and Iraq [9].

Gambling and Internet addiction

Virtual reality seems to be a useful complement to cognitive-behavioral therapy in the treatment of various types of addiction, especially those involving the use of computers; in these cases, a real de-conditioning is performed.

The psychoses

As is well known, psychoses are mental illnesses characterized by delusions, hallucinations, behavioral disturbances, language disorders, and reduced or altered expression of emotions.

In this type of disturbance, virtual reality devices, and in particular, the models of Avatars (i.e., images or digital graphic representations in which the subject recognizes himself and identifies with them, sometimes establishing a particular relationship, feeling them friendly and protective parts of himself) have given surprising and unexpected results; for example, appropriately designed avatars have reduced auditory hallucinations in some psychotic patients [4]. The result is of great interest because it demonstrates that in still unknown ways, virtual reality can positively interfere in serious neuro-psycho-pathological processes such as psychosis.

Reflections

The potential of Virtual/Augmented Reality is enormous: the availability in advanced countries of software and hardware dedicated to them will allow enormous progress but can also, once again, increase the power and wealth of a small elite, leaving the possibilities to the great masses to be uncritical consumers (obviously paying) or marginal subjects. Their market is estimated to be worth nearly $90 billion by 2023 [8–10]

Both technologies are currently in an early stage of development. In particular, there is a lack of studies showing their incremental efficacy in the clinical setting.

It is therefore essential to develop an adequate knowledge of these very important innovations, always remembering that man is the measure of things, and to try, as far as possible, to contribute to their highly qualified management but also democratic and "on a human scale."

References

[1] Riva G, Wiederhold B. et al. Neuroscience of virtual reality: from virtual exposure to embodied medicine cyberpsychology. Behavior Soc Netw Available from: http://doi.org/10.1089/cyber.2017.29099.gri.

[2] Chiamulera C. Virtual reality: virtual experiences between technology and brain. Hachette Edit. Milan, [•]; 2018.
[3] Hunter ECM, Carlton J, David AS. Depersonalisation and derealisation: assessment and management. BMJ 2017;356: j745. Available from: http://doi.org/10.1136/bmj.j745 (Published 2017 March 23).
[4] Tom KJ, Craig, Mar Rus-Calafell M. et al. AVATAR therapy for auditory verbal hallucinations in people with psychosis: a single-blind, randomized controlled trial thelancet.com/psychiatry on November 29, 2017. Available from: http://doi.org/10.1016/S2215-0366(17)30427-3
[5] Tang R, et al. Augmented reality technology for preoperative planning and intraoperative navigation during hepatobiliary surgery: a review of current methods. Hepatob Pancreat Dis Int 2018;17:101–12.
[6] Cho K, et al. Craniofacial surgical planning with augmented reality accuracy of linear 3D cephalometryc measurement on 3D holograms. Plast Reconstr Surg Glob Open 2017;5(suppl):204.
[7] Best J. How virtual reality is changing medical practice: "Doctors want to use this to give better patient outcomes". BMJ 2019;364. Available from: https://doi.org/10.1136/bmj.k5419.
[8] Ramachandran V. The Tell-Tale Brain 2011 Trad Italiana: The man who thought he was dead. Mondadori ed. Milan; 2013.
[9] Rizzo A, Jarrell Pair J. et al. Development of a VR therapy application for Iraq War Veterans with PTSD. University of Southern California Institute for Creative Technologies 13274 Fiji Way, Marina del Rey, CA. 90292.
[10] Augmented Reality and Virtual Reality Market by Offering (Hardware & Software), Device Type (HMD, HUD, Handheld Device, Gesture Tracking). Application (Enterprise, Consumer, Commercial, Healthcare, Automotive), and Geography - Global Forecast to 2023 augmented. https://www.marketsandmarkets.com/Market-Reports/augmented-reality-virtual-reality-market-1185.html.

CHAPTER

15

The robotic assistance system

Summary

For decades, computer experts have been discussing the great potential of robots as collaborators of human beings. The huge recent advances in artificial intelligence (AI) have made it possible to create robots that are truly capable of responding to various care needs. Let us see what they can do.

A robotic system can be defined as an artificial intelligence (AI) capable of interacting in the physical world, an embodied AI or as a *"smartphone with hands,"* to use the analogy of Giorgio Metta, deputy scientific director of the Italian Institute of Technology of Genoa.

The term robot currently has several meanings and is functional for multiple applications in industrial, military, or rescue operations. In the medical context, in addition to the automata used in the diagnostic, surgical, therapeutic, and rehabilitative fields, an emerging sector is that of the so-called "assistance robots," machines capable of carrying out tasks related to assistance in the physical or emotional field.

Designed to provide help and social interaction to people in everyday life, as well as cognitive support, training, and support for operators, robots can help with shopping, welcome in waiting rooms or clerks in department stores, help with doing homework, housework, and becoming "friends" with the elderly.

Care robots can be divided into three categories: robots with monitoring functions, so-called "assistive" robots, and robots with companionship and therapy functions [1]. Here we describe the robots belonging to the third category, proposed for companionship and carrying out physical, social, and leisure activities, and to creating an artificial form of *pet therapy*.

AI in Clinical Practice
DOI: https://doi.org/10.1016/B978-0-443-14054-9.00003-X

Examples of robots designed to assist older people include Paro, (comPAnion Robot), designed in Japan, with the appearance of a baby seal, able to interact independently with people, has a wide range of behaviors, such as example can move the fins, raise the head and blink, so as to be expressive. It recognizes its own name but can learn a new one in order to be "renamed" by its users. It understands about 500 words in English, even more in Japanese. It has subcutaneous sensors that allow it to sense if and how it is touched, whether softly or aggressively. It is covered in soft white fur, which gives it a plush appearance that makes people want to touch or pick it up. It does not have to be educated about cleanliness or fed. Just reload it by connecting it to the mains by means of a soother-shaped socket that is inserted into its mouth.

The actual impact on assisted people and their families

Several studies have shown a positive effect of social robots on the mental and physical health of the elderly. In particular, they would improve the ability to manage stress and would also have a positive effect on mood [2]. The studies carried out in this area are also of low quality, observational, on small numbers, and without a control group. Positive short-term changes are generally described, especially with *pet therapy* with "real" animals, without the possibility of highlighting which component of the intervention is the cause, for example, the increase in social relationships or the contribution of Announcements. There is certainly an important placebo effect. Efficacy studies are, however, difficult to carry out; just think of double-blind use. Pharmaceutical companies are also certainly not interested. Innovative approaches are probably needed, for example, qualitative studies with rigorous methodology associated with quantitative studies that can summarize the results of multiple works [3].

The incessant technological developments have not led to an industrial-scale production, but the service robot market is nevertheless constantly expanding, and significant growth is expected in the coming years. In Italy, in particular, there is a lack of political support and industrial investment so that handcrafted prototypes can become mass products [4].

Introduced for the first time in a 1920 theatrical drama, R.U.R. (Universal Robots of Rossum), by Czech author Karel Capek, the word derives from the ancient Slavic robota, which indicates the servile and toil that the peasants had to perform to the owner's landowners. For this linguistic origin the term must be pronounced ròbot and nonrobò in French (from Henin S. AI - Artificial intelligence between nightmare and dream, Hoepli, Milan 2019).

Homo roboticus and sapiens sapiens

The American Sherry Turkle, sociologist, psychologist, and technology expert, defines our era as a robotic moment, not because companion robots are widespread in our reality, but in reference to the state of emotional availability of many people who are in favor of seriously considering robots even as friends, confidants and even loving partners [5].

Social robots, designed to satisfy the intrinsic needs of human beings for "affective" bonds, do not perceive what humans perceive; they cannot "feel" anything, and they cannot experience the sensation of social interaction. Robotic empathy is, therefore, not possible. Moreover, according to some philosophers, such as Luisa Damiano and Paul Dumouchel, internal states are not indispensable for an emotional relationship, which robots do not have and cannot have, but it is only necessary that the artificial agent is able to modify its own behavior in function of the emotional expressions of the social partners, what the aforementioned authors call "artificial empathy." [6]

Is it still acceptable for fragile and vulnerable people to become attached to robotic agents that basically pretend to have emotions but do not have any? According to Sherry Turkle, the robot has considerable therapeutic potential, as it allows people to be cured by allowing them to offer, even to an effective automaton, the comfort that they themselves really need. The mere representation of the emotional bond would be a sufficient bond, moreover, toward a behavior, certainly not a feeling. Turkle also points out that the emotional bonds fostered by robots are deceptive and that people can lose the desire and capacity for human relationships.

According to other experts, referring to the "material" basis of thought, it would be possible that robots could express real emotions. Sadness could be achieved, for example, by setting a specific code. This would be similar to that of humans, being also basically a number, a number of the quantities of neurochemicals present in the brain. Why should a car number be less authentic than a human's [7]? Could not a robot's emotions be made of the same biochemistry as that of a human? Is not the mind made of matter? Should we, therefore, accept/conceptualize a new category of emotions, different from humans even if probably genuine and authentic? [8] A disturbing question, also because an excessive sharing of feelings with robots can, for example, accustom us to a reduced range of emotions, with the risk of lowering expectations regarding relationships with real people. It is also true that human beings disappoint, but robots do not. These do not abandon, while friends and family members often do not understand (or pretend not to understand) the needs of the elderly, more or less implicitly often considered a burden. According to some, a social robot as a companion-assistant would be better than no care, and in any case, better than a person in some cases distracted, listless, in an atmosphere of icy indifference.

Surely a robot is not currently able to understand if an elderly person is worried or sad or if he wants to die: but how many professionals are really able to do it? If assistance is further standardized, reduced to a predefined script, and performed mechanically, it is perhaps easier to accept an auxiliary *homo robot* rather than a *sapiens sapiens*.

On the other hand, the either/or choice that provides only two alternatives is not acceptable: either companion robots or condemnation of loneliness. There may be intermediate solutions, robotic helpers capable of doing the humblest and most repetitive jobs, for example, machines that can turn weakened or paralyzed patients in bed so that they can be washed. In this case, the robot would not carry out an autonomous assistance activity but would be an extension of human beings, who could thus have more time to deal with the more personal and emotional aspects. Moreover, relationships are based on the time invested, even by carrying out the most trivial activities, which confirm our ability to love and take care of others. For example, mealtime can be an important social and emotional event.

Reflections

The debate on assistive robotics (and socially in general) is extremely current, process and product technologies are rapidly developing, and the future scenarios of robotic assistance find ample space also in the mass media.

Care robots, increasingly sophisticated and developed than those currently available, can enhance people's autonomy.

It is, however, difficult to accept the concept of robots as assistance personnel, caregivers, and e-persons (electronic people), since the care of people is based, above all, on relationships on a personal, social, and emotional level, and its fundamental aspects are nonverbal language, thoughtfulness, attention, closeness, understanding. Human characteristics, simply but exclusively human, are the only ones that allow us to build real care communities.

From this point of view, a robot that provides assistance appears to be something inhuman, deceptive, and inappropriate: the automaton does not prove anything of what we do. As Roberto Cingolani, scientific director of the Italian Institute of Technology in Genoa, affirms: "*he lacks the biochemistry of life and therefore the irrational, irreproducible part, capable of altruistic impulses or aggression which, instead, characterizes man.*" [9]

It is, therefore, time for health professionals to participate in the ethical reflections relating to the conditions of use of assistive robots, to the questions of responsibility we have toward each other, and to the founding principles as human beings. We must ask ourselves whether it is right to accept to assign artificial companions to the elderly, used as remedies for the isolation of old age and sometimes even for our feelings of guilt.

We must reflect: elderly people who are not self-sufficient are often considered subjects unable to express their realities, which are, therefore, established by strangers: from here to deciding that these people would not need assistance from human beings, the step is short; it could follow that the companionship/assistance of one's fellowmen will end up being granted only to the wealthy and to those who do not have physical and mental problems [10].

The theory of the uncanny *valley*, that is the uncanny valley

The psychological phenomenon, described by the Japanese robotics teacher Mori Masahiro [11], according to which the feeling of familiarity that we can feel toward anthropomorphic machines grows the more the similarity with us increases, but only up to a certain point, after which we witness a sort of emotional rejection, signaled by a sharp bending in the curve that represents, in a hypothetical graph, the attitude toward these artificial entities. In practice, the robot, to have to do with us without causing discomfort, must look like us but not too much!

References

[1] Van Aerschot L, Parviainen J. Will robots take care of us? https://www.ingenere.it/articoli/saranno-robot-prendersi-cura-di-noi

[2] http://www.parorobots.com/whitepapers.asp

[3] Burton A. Dolphins, dogs, and robot seals for the treatment of neurological disease. Lancet 2013;12:851–2.

[4] Caldelli V. Lisa and Diago, domestic robots. https://wsimag.com/it/scienza-e-tecnologia/23646-lisa-e-diago-robot-domestici

[5] Turkle S. Together but alone. Because we expect more and more from technology and less and less from others. Turin: Editions Code; 2012.

[6] Dumouchel P, Damiano L. Living with robots. Essay on artificial empathy. Milan: Raffaello Cortina Publisher; 2019.
[7] Brooks R, Quoted in MIT: creating a robot so alive you feel bad about switching it off—A galaxy classic. Dly Galaxy, December 24, 2009.
[8] Brezeal C, Brooks R. Robot emotion: a functional perspective. In: Fellous Jean-Marc, Arbib Michael, editors. Who needs emotions: the brain meets the robot. Cambridge: MIT press; 2005.
[9] Cingolani R. The other species. Bologna: Il Mulino; 2019.
[10] Collecchia G. Assistance robotics. Recent Prog Med 2019;110(10):473–5.
[11] Mori M. The uncanny valley. Energy 1970;7(4):33–5.

CHAPTER

16

Will "amplifying technology" really expand our knowledge? what is amplifying technology?

"Amplifying technology" is the term used to describe a series of devices that amplify human perceptions, generally improving their performance. Among the most well-known and effective devices we remember those used by fighter aircraft pilots!! Other amplifying devices can be used for rehabilitation purposes for example in neurological patients.

Does perceiving more always mean knowing and understanding more?

The companies that produce amplified devices claim that thanks to them, we can perceive more and better, and therefore, we will know more and understand more and more things: this hypothesis is taken up and confirmed by various popular magazines.

Neuroscience, however, teaches us that the relationship between perception and knowledge is much more complex [1]. Each perceived data is integrated with many other data, compared with the patterns we have learned over several years, stored and recorded on a priority scale (see also table on sensors and virtual reality).

So the amplification of our perceptions will not always have positive consequences for us: it could be insignificant or even negative.

The concept was recently well expressed by Federico Cabitza, Professor of computer science, at the 4words 2018 [2] conference; he warned that the tendency to observe an analogy between sensory amplification and cognitive enhancement is misleading and dangerous.

In fact, we tend to think that the same technology that strengthens and amplifies our senses (think of the stethoscope, the microscope,

DOI: https://doi.org/10.1016/B978-0-443-14054-9.00014-4

the X-rays) can easily lead to equally increasing our interpretative and cognitive abilities.

In fact, the scientist recalls that Galileo Galilei, in 1610, in *Sidereus Nuncius*, having observed and then described the visible side of the Moon, thanks to a first rudimentary telescope, concluded that observation is one thing, interpretation of this is another, which could be observed, also by means of an available technology capable of amplifying the senses.

The author also points out that, for example, the same radiological image can be interpreted by experts in a completely different way.

"The point is that the decisive element lies not so much in the sense of sight, enhanced by technology, but in the correct interpretation of the phenomenon observed, in understanding what is being observed and inserting it into a framework of overall sense making," concludes Cabitza.

With regard to amplification, it is rather an alliance between the machine and man, to which each part contributes by making the best of itself available: the ability to calculate, search, and find information within an enormous amount of the given machine; intuition, creativity, communication of the doctor or the human being. At the moment, only man is able to grasp the extent of the variability and the ambiguity of the context and to process these complex data with great accuracy.

References

[1] Eric R, Kandel J, Schwartz H, et al. Principles of neuroscience Casa Editrice Ambrosiana Milano; 2019.

[2] Cabitza F. 4 words 2018. The words of innovations in healthcare. Suppl Recenti Prog Med 2018;109:4 April.

CHAPTER

17

Diagnosis by images. The recognition of images in humans and in artificial intelligence systems

There are important differences between the visual processes of humans and the systems of interpretation of the images of artificial intelligence. We summarize them briefly to begin to know their limits and potential.

Humans

The fundamental principle of human vision is that "Seeing is remembering and comparing" [1].

The images are perceived by the retina and transmitted to various nerve centers that organize them in the way described by the perceptual psychology of Gestalt: figure-background, proximity-distance, continuity-discontinuity, similarity-diversity, natural forms "good forms" and unnatural, closing and opening of lines, surfaces, spaces, etc.

Thus we arrive at an overview that our nerve centers will catalog according to models and maps built by us during our life. The process is fast and largely unaware.

A singular visual phenomenon, which is the basis of human vision and which is not present in AI systems, is the resolution of the "Inverse Problem of Optics."

Different images have the same retinal projection but appear quite distinct thanks to our nervous system which processes them with very efficient mechanisms even if largely unknown (1).

AI in Clinical Practice
DOI: https://doi.org/10.1016/B978-0-443-14054-9.00017-X

The "inverse problem," however, exposes us to embarrassing visual illusions, many of which are masterfully illustrated by the great artist Cornelis Escher.

The visual illusions are, however, highlighted and corrected by our wonderful brain through a simple expedient, an instinctive ingenious mechanism, the second glance: instinct (or perhaps the Socratic daimon?). In a split second, it tells us that it is not possible and that we have to review the image.

Superintelligent machines do not possess this faculty unless any unexpected and/or paradoxical situation is explicitly foreseen by programmers.

Digital images

The digital image is made up of optical points (pixels), each of which is associated with numerical coordinates recorded in the computer's memory: the computer uses algorithms that form the image from numerical coordinates.

The information extracted can allow different levels of precision: low level (such as statistics on the presence of various shades of gray or colors, on sudden changes in brightness, etc.); intermediate level (characteristics related to image regions and relationships between regions) or high level (object recognition).

While for the low and medium levels, software and algorithms provide us with excellent performances, for the higher levels, the interpretative ones, there are still great difficulties: an interesting Canadian study has shown how the simple introduction of a fake elephant in a room has completely disoriented the artificial intelligence system that was no longer able to recognize even the previously identified objects [2].

This serious limitation is a major concern of auto-driving system experts, who are also bewildered by events not explicitly predicted by the algorithms. For example, a hen crossing the road completely disorients the visual recognition systems.

Visual recognition systems

If we are to believe the companies that make them, visual recognition systems are almost infallible. The fact is that sensational errors are described and documented: from a poor man beaten by the police in Denver (USA), imprisoned and acquitted after a year of detention, when the error of the algorithm was recognized [3], to an intelligent child of 10 years that he violated the sophisticated visual recognition system of a prized Apple iPhone, and to prove that he could at his discretion reviolate it, he published the video on YouTube [4].

Images in digital diagnosis

In the last decade, with the use of neural networks and deep learning, digital diagnostics has made an enormous qualitative leap, linked not only to the enormous potential of these intelligent systems but also to the fact that the creators of algorithms are generally able to provide very precise and above all comprehensive information.

As we will read in the following chapters, dedicated to the application of Artificial Intelligence to diagnostics, if the system knows exactly what it has to do, it does it well and much earlier than humans.

The open problem is, even for superintelligence, the unexpected and/or the exception.

References

[1] Purves D, Brannon E, et al. Cognit Neurosci. Zanichelli Ed., Bologna; 2014, p. 133.
[2] Rosenfeld A, Richard Zeme R, Tsotsos J, The elephant in the room. https://www.researchgate.net/publication/326995550_The_Elephant_in_the_Room; 2018.
[3] Fry H. Hello World. Bollati Editore, Milan; 2018, p. 156.
[4] Watch a 10-Year-Old's Face Unlock His Mom's iPhone X. https://www.wired.com/story/10-year-old-face-id-unlocks-mothers-iphone-x/

CHAPTER

18

Machine learning, deep learning, and clinical studies

Machine Learning (ML) and Deep Learning (DL) produce results (output) that are strictly related to the data introduced (input), their structural characteristics, and the training methods used for the machines.

The processing speed, the enormous quantity of data produced, and sometimes the surprising originality of the results can easily lead to an overestimation of these devices, erroneously considered infallible.

Furthermore, the natural tendency of the human mind to obtain the best possible result with the least effort leads us to accept uncritically, as "oracular wisdom," the conclusions of artificial intelligence.

The JAMA, the Journal of the American Medical Association, an authoritative American journal [1–3], and the BMJ, an important English journal [4,5], provide us with precious methodological indications that help us to evaluate the medical studies carried out with methods of artificial intelligence.

Study validity

What to evaluate?

1. Are the methodologies and procedures used in the study clearly described? Are they correct?
2. Were the procedures and data also checked by human experts or only by artificial intelligence devices?
3. Are there references and comparisons with studies already carried out on the subject and considered reliable?
4. Q) Is the machine training algorithm controllable, and is it consistent with the health objectives proposed by the study?

AI in Clinical Practice
DOI: https://doi.org/10.1016/B978-0-443-14054-9.00023-5

Results

1. Does the study modify and enrich in some way our knowledge on that subject?
2. Is the study reproducible under the same conditions, and is it transferable in different realities and patient populations different from those studied?
3. In the event that it brings new knowledge, does the study include verification and falsifiability criteria, or are there events that, if they occur, can deny it? (See Note)
4. In the event that the results of the study are verifiable, reliable, and reproducible, its applicability in the specific reality in which we operate and in relation to the patient population we follow must be further evaluated.

Notes

The falsifiability criterion, proposed by Karl Popper [6] as the scientific validation criterion of all new acquisitions, is an important control tool also with respect to the results provided by Deep Learning devices: if, for example, the artificial intelligence system estimates that a new drug active on the coronavirus can eradicate it in a percentage between 60% and 80% of cases, it will certainly be possible in a reasonable period of time to confirm or deny the initial hypothesis.

If, on the other hand, the same artificial intelligence system estimates the efficacy of a drug on the basis of surrogate end-points (e.g., reduction in blood sugar and/or cholesterolemia as a substitute for a decrease in cardiovascular complications), we should accept these conclusions very cautiously, as they are not based on irrefutable evidence: to those who have doubts about it, we remember the unhappy experiences of pioglitazone for the treatment of diabetes and above all of cerivastatin as a cholesterol lowering drug.

It is, therefore, of great importance, in the face of the multiplication of research based on artificial intelligence, to apply a method of reading and evaluation inspired by the same criteria of rigorous criticism that our imperfect mind has used for centuries for thousands of studies and experiments, which have guaranteed to humanity an imperfect but still better world.

References

[1] Yun L, Po-Hsuan CC, et al. How to read articles that use machine learning users' guides to the medical literature. JAMA 2019;322(18):1806–16. Available from: https://doi.org/10.1001/jama.2019.16489.

[2] Derek CA, Randomized clinical trials of artificial intelligence. JAMA 2020;323 (11):1043–5. Available from: https://doi.org/10.1001/jama.2020.1039.
[3] Michael EM, Danielle W, et al. Artificial intelligence in health care: a report from the National Academy of Medicine. JAMA 2020;323(6):509–10. Available from: https://doi.org/10.1001/jama.2019.21579.
[4] Sebastian V, Bilal AM, et al. Machine learning and artificial intelligence research for patient benefit: 20 critical questions on transparency, replicability, ethics, and effectiveness. BMJ 2020;368:l6927. Available from: https://doi.org/10.1136/bmj.l6927.
[5] Myura N, Yang C, et al. Artificial intelligence versus clinicians: systematic review of design, reporting standards, and claims of deep learning studies. BMJ 2020;368:m689. Available from: https://doi.org/10.1136/bmj.m689.
[6] Karl P, Logic of scientific discovery Einaudi Edit. Turin pg.66% 83%; 2010.

CHAPTER

19

At the end of the journey… Conclusive reflections

"……*Philosophy always comes too late. As a thought in the world, it appears for the first time, after reality has completed its process of formation. When philosophy paints in chiaroscuro, then an aspect of life has aged, and, from the chiaroscuro it does not allow itself to be rejuvenated, but only recognized: Minerva's owl begins its flight at dusk*" (GWF Hegel [1]).

The introduction in the health sector of intelligent systems, capable of learning and deciding, opens up exciting new frontiers but at the same time radically changes the relationship between man and technology, which is increasingly becoming a sort of alien species, even if not yet hostile.

Moreover, the future is not determined, and it is not something that happens. We create it. It is possible to intervene on its development, to try to reintroduce into the culture of medicine an explicit dialectic, a comparison of thoughts, methods, and objectives necessary to enter into the merits of the processes, of the management of projects, of the defense of citizens/patients, avoiding a sort of *PowerPoint* logic, characterized by "given" communications, *fashion* presentations of realities considered certain and definitive in an optimistic way [2].

AI is changing the cultural paradigm of medicine, the scope of which alone could legitimize its very existence. Algorithmic applications and related analytical skills could become increasingly indispensable, if well integrated with the activities of health professionals, to provide clinically important answers, especially in highly complex contexts, and allow doctors to have more time to take load the care needs of their patients.

Algorithms present us with a possible paradigm shift in predictive power. We no longer have to observe a complex system for years and build empirical models, but we can have algorithmic constructions to predict the future [3].

DOI: https://doi.org/10.1016/B978-0-443-14054-9.00015-6

The mind of man, even the brightest, is fallible, unable to memorize excessive amounts of information and to recall and analyze the enormous amount of data necessary for the care of people.

Moreover, data are not values, and any intervention based on them must be endowed with meaning validated, also considering the frequent contradiction of the knowledge provided by the literature.

The AI will be essentially useful as a complementary for the doctor, who will be able to delegate the calculations and operations on the data to the machines but keep the interpretation of complex phenomena and the consequent possible solutions for himself.

Doctors must play a guiding, supervising, and monitoring role (being *"in the loop"* rather than *"out of the loop"*), using their intelligence ("the courage to use our intelligence," to use the famous phrase of Kant) and the skills that make them superior to machines, in particular abstraction, intuition, flexibility, and empathy, the so-called "soft skills," aspects of the profession that an algorithm will never be able to reproduce. Humans must exercise a conservative and constructively critical approach towards machines, highlighting their enormous potential, often uncritically emphasized for commercial reasons, but also their limitations (and possible threats, such as the sci-fi dystopia of machines in power!).

To exercise this control, human supervision must be of high quality and therefore requires high-level training courses, adequately funded. It is certainly necessary to raise the awareness of all health personnel to acquire IT skills, digital systems, and biostatistics. In an ideal world, physicians should know the basics of algorithm design, how to obtain datasets for outputs, and have the skills to understand the limitations of algorithms.

The best results are therefore expected when AI works in support of health personnel, "second set of eyes," a way of cultural integration between humans and *smart machines*, avoiding emphasizing disputes, basically irrelevant, on which cognitive system, human or artificial, is more "intelligent." Quoting A. Verghese, *"clinicians should seek an alliance in which machines predict (with significantly greater accuracy) and humans explain, decide and act"* [4].

AI systems must be considered a tool, such as a microscope, the stethoscope, and the electrocardiograph, developed over time to make up for the limited perceptive capacity of doctors.

As stated by E. Topol, in medicine, AI should not exceed level 3 of the scale used to classify the level of autonomy of cars, a level of partial autonomy conditioned by human supervision [5].

The writer considers a different cultural approach to be fundamental, capable of composing the relationship between technologies and care needs. A modality that could be defined as "vectorial," rather than punctual as often happens: a change of direction, perspectives, health objectives, and the consequent reference variables: equality, care needs, access to information and care/services, health planning, continuity of care. Overall, a redefinition of priorities, quality, and equity of care, giving space to people's requests to provide answers based on a global approach, focused on the identification and sharing of values, meaning, and objectives.

Robots, for example, invite us to reflect on how we want to be and what kind of people we want to become since we are launching ourselves into ever more intimate relationships with machines and less and less towards the rights and respect of real people. Perhaps it is wrong to remain fixed to predefined interpretative categories and ethical codes, anchored to the past but (precisely for this reason) constitutive of human identity. The questions are complex and currently have no definite answers. Should we adapt to the artificial evolution of the unprecedented ethical issues emerging from the nascent mixed human-robot social ecologies? Should we accept the possibility of synthetic emotions, representations of our own, coming from objects made by us, accept/surrender therefore to a disturbing "synthetic" ethics [6]?

In the real world, the limits of the current health system are known to all, not only in our reality; limits, and at the same time, waste of resources, possible diagnostic errors, insufficient time dedicated to the doctor-patient relationship, inequality of access. The same explosion of science and technology taking place worldwide to tackle the ongoing COVID-9 pandemic risks deepening the gap between developed and backward peoples and between rich and poor within the advanced countries themselves.

With AI algorithms, it may be possible to improve and enhance health services. For example, if currently, with a CT scan we can do 50 scans a week, applying the AI, able to identify any pathologies and in general situations of uncertainty to be submitted to the doctor's judgment, we could reach 500, increasing the time available to the doctors for other tasks. The loss of jobs is, therefore, not the real issue but the training of doctors so that they are able to offer added value [7].

Algorithms can strengthen clinical governance and optimize resources, such as addressing the impending shortage of professionals and theoretically achieving a reduction in healthcare costs, but interdisciplinary collaboration is essential to ensure that the goals of programmers coincide with those of clinicians. Before implementing AI systems in clinical practice, possible errors in algorithms, inefficiencies, and costs must also be reduced.

This requires rigorous studies, published in peer-reviewed journals, of real-world clinical validation against important clinical outcomes, such as reduction in morbidity/mortality or improvement in patients' quality of life, but also the level of satisfaction, both physicians and patients, in the new relational context.

Studies should, in particular, include comparisons in terms of clinical outcomes between groups of physicians who use AI systems and others who do not use such technological support. Finally, postmarketing monitoring of the systems approved by regulatory bodies must be provided, similar to that commonly used for drugs and devices, with the aim of highlighting, in the long term, possible unexpected consequences deriving from their practical application in contexts other than experimental ones [8].

The added value of AI and digitalization in patient care must be effectively demonstrated in clinical practice, and therefore, recognized and remunerated by the Health Systems to allow *healthcare* companies to invest. Otherwise, the implementation of the technology will represent a waste of resources and the price to be paid will be the loss of time to devote to the sick to power the systems.

It is legitimate/necessary to find alternative solutions to health systems that are collapsing, but technology, E. Sadin's "algorithmic orientation of behavior," cannot replace or even interfere in the doctor-patient relationship. Health policies should support and sustain the medical culture based on a trustworthy and basically demedicalizing approach to the person, finance them adequately, and relieve them of excessive bureaucratic and administrative burdens, avoiding encouraging initiatives that can lead to unproductive costs in terms of health, often deriving from an improper use of the same technologies.

The history of our civilization, however, can help us to understand and choose: from the invention of the wheel to the discovery of atomic physics, the human species only progressed when it has kept technology under control and used wisely and prudently.

When, on the other hand, it was deluded that a free development of technology would allow great leaps forward, it found itself running great risks: it happened so for chemistry, but even more so for atomic energy.

History is repeating itself with artificial intelligence; as long as we control it, we can only have great benefits; let us not forget that the most revolutionary discoveries in the history of human thought occurred when the great scientists allowed themselves to be influenced by imagination and emotions: to cite just a few examples, by observing the puffs of smoke from his pipe relaxed, the chemist Kekulè had the original intuition that benzene did not have a linear but a ring structure (a completely innovative idea for the time). Einstein, playing the violin, flew with his imagination beyond classical physics, and Schrödinger, playing with his cat, concretized in the famous "cat paradox," revolutionary concepts of quantum physics.

If instead, we free the machines from our continuous control and let ourselves be guided not by our fantasies and emotions but by their uncontrollable algorithms, we will be gradually expropriated of our creativity and we will end up hypnotically following the sound of a new Pied Piper from time to time.

It is ironic that precisely when time in clinical practice is increasingly limited, a profound reflection on the possible effects of the transformation taking place is indispensable, a thought of the digitalization of the world, a renewed humanistic awareness, in terms of acceptance by the carers and of all operators, of changes in a professional role, of relationship with the patient, of indispensable training needs.

It is necessary to flee as much from apocalyptic visions of the future by trusting in human reason as from an accentuated enthusiastic attitude towards the potential of technology by recognizing its limits.

We need to identify a third way, what Julian NIda-Rumelin and Nathalie Weidenfield define digital humanism [9].

The alternative is to lose the game, or in any case, to get to know the magnitude of the phenomenon too late, once things have been done, such as Minerva's owl, which arrives when the reality is "good and done."

Paraphrasing H. Fry, in the age of algorithms, the human being, guided by his own intelligence, has never been so important [10].

Unfortunately, the ongoing pandemic reminds us that "the"; computing "capacity of collective intelligence expressed by viral populations, to find adaptive solutions to human defensive responses, still far exceeds the potential of any alleged other collective intelligence, human or artificial" [11].

References

[1] Hegel GWF. Outlines of philosophy of law. Bompiani 2006.

[2] Collecchia G. Digital medicine: the empire of sensors. IsF 2018;3:21–8.

[3] Vespignani A, Rijtano R. The algorithm and the oracle. Il Saggiatore, Milan.

[4] Verghese A, et al. What this computer needs is a physician. Humanism and artificial intelligence. JAMA 2017;319:19–20.

[5] Topol E. High-performance medicine: the convergence of human and artificial intelligence. Nat Med 2019;25:44–56.

[6] Brooks R. Quoted in MIT: creating a robot so alive you feel bad about switching it off—a galaxy classic. Dly Galaxy 2009; December 24.

[7] Rajan R. Il Sole 24 ore, 12 January 2020.

[8] Rasoini R, Alderighi C. On the human attitude towards technological innovations: from the mechanical turk of the 18th century to contemporary systems of artificial intelligence in medicine. Toscana Medica 2019;5:24–6.

[9] Nida-Rumelin J, Weidenfield N. Digital humanism. An ethics for the age of artificial intelligence. Milano: Franco Angeli; 2018.

[10] Fry H. Hello World-Being human in the age of machines. Turin: Bollati Boringhieri; 2018.

[11] Corbellini G. Pathocenosis of Covid-19: a tribute to Mirko Grmek. https://www.scienzainrete.it/article/patocenosi-di-covid-19-tributo-mirko-grmek/gilberto-corbellini/2020-05-09?utm_source = phplist1435&utm_medium = email&utm_content = HTML&utm_campaign = Cronache + della + ricerca + %23134.

CHAPTER

20

ChatGPT is beginning to change medicine

ChatGPT (CGPT) stands for "Chat Generative Pretrained Transformer," that is, "conversation-generating pretrained transformer." It is a chatbot developed by OpenAI, the US-based company founded by Elon Musk, specializing in conversation with human users based on artificial intelligence (large language model system, i.e., large-scale language model). Basically, it is an advanced form of conversational artificial intelligence, a powerful tool to statistically sequence natural human language with several potential applications: answering questions, correcting texts, making summaries, word processing, writing poetry, translating, making tables, and proposing diagnoses and therapies in medical fields.

It relies solely on language and is trained to predict the probability of a given sequence of words based on the context of those that precede it. It chooses the sentences that are most likely to make logical sense (attributed by us) with respect to what it has studied.

It is basically based on the probability that after a word there is another word; after a sentence, there is another sentence; after a question, there is an answer. It determines these probabilities by having received as input, in "pretrained" training, billions of sentences available on the Internet and valuing hundreds/thousands of billions of parameters!

Basically, it would be nothing more than a stochastic parrot[1] repeating sentences that WE have written down somewhere (Internet, books, social) without understanding their meaning in any way.[2]

1 accoto-harvard-business-review-italia-marzo-2023.pdf.

2 On the Dangers of Stochastic Parrots: Can Language Models Be Too Big? 1F99C (acm. org).

DOI: https://doi.org/10.1016/B978-0-443-14054-9.00031-4

Launched on November 22, 2022, ChatGPT has reached 100 million users in just two months. For now, its training is based on a dataset updated to 2021.

How to use ChatGPT in general medical practice?

We asked ChatGPT how it can be used as a support tool in general medical practice in various ways. Here is its answer:

1. **Answers to general questions:**
   You can use ChatGPT to get answers to basic medical questions, such as information about symptoms, causes of common diseases, risk factors, treatment options, and preventive measures; however, it is important to keep in mind that ChatGPT is not a substitute for a doctor and its answers should be considered only as basic information.
2. **Patient education:**
   ChatGPT can be used to provide educational information to patients on a wide range of topics, such as healthy lifestyle, diabetes management, cardiovascular disease prevention, the importance of vaccinations, and more. This can help patients better understand their medical conditions and adopt healthy behaviors.
3. **Support in diagnosis:**
   ChatGPT can be used as a support tool to help physicians gather information about patient's medical history. For example, ChatGPT can ask standardized questions to obtain information about symptoms, their duration, risk factors, family history, and more. This information can be useful in establishing a preliminary diagnosis or identifying potential diagnostic guidelines to follow.
4. **Clinical decision assistance:**
   ChatGPT can help clinicians make informed clinical decisions by providing information based on clinical guidelines, treatment suggestions, or disease management considerations; however, it is important to emphasize that final decisions should always be made by a qualified physician based on a comprehensive evaluation of the patient.
5. **Support in medical research:**
   ChatGPT can be used as a tool to quickly access up-to-date medical information and scientific literature. You can ask specific questions about medical research, the latest studies, or available treatment options for a given condition.

ChatGPT: How does it work?

The great success of ChatGPT is related to its ability to understand human language and respond appropriately to the questions posed. Knowledge of how CGPT analyzes the questions it is asked allows us to understand its great potential but also to intuit some of its limitations.[3,4]

1. The first stage of natural learning processing (NPL) analysis is the fragmentation or "Tokenization" of each passage that is broken down into sentences.
2. The second stage is "Parts of Speech (POS)" labeling, in which each sentence is examined, and the various words (nouns, pronouns, verbs, adjectives, and so on) are identified and labeled.
3. The third stage is named entity recognition (NER) of the key entities in the sentence. Artificial intelligence identifies the semantic sets to which nouns, verbs, adjectives, and so on belong. For example, proper nouns are distinguished from common nouns, active verbs from passive verbs, etc.
4. The fourth stage is the analysis of the relationships and dependencies among the various constituents of the phrase "Parsing". It allows reconstructing the links between the various components of the sentences and evaluating the strength of these links.
5. Semantic analysis is the final part of NPL, which is the reconstruction of the meaning of individual sentences and the overall meaning of the passage or "prompt."

Knowledge of the mechanisms of language analysis allows us to immediately realize an important limitation of AI systems; while they are almost perfect from a logical point of view, they do not possess an innate and instinctive ability of humans that plays a very important role in many contexts; "good sense," or the "mental disposition to balance and measure that manifests itself in the aptitude to make sound decisions by adapting norms and principles to the problems that arise from time to time" [1]. In other words, although CGPT always performs context evaluation, the purely human ability to go beyond pure literal meaning and use our intuition and empathy to understand others is not currently reproducible in artificial intelligence.

Numerous scholars have evaluated and continue to evaluate the potential and limitations of CGPT. We report some recent research evaluating its important applications in health care.

3 https://hiv.net/aiologyhandbook.

4 https://www.elementsofai.com/.

A group of researchers at the University of San Diego in California asked CGPT 195 medical questions comparing its answers with those already given by verified, experienced physicians [2]. The result was reassuring; CGPT provided answers that were not only as correct and adequate as those of the physician, but also often clearly precise and more comprehensive; however, there are definite limitations of the study, since the questions were derived from a social medial platform and the answers were provided by physicians not actually involved in the care of the patient asking questions.

A similar research group at the same university asked CGPT 23 important questions in four areas of public health (addiction problems, mental problems, interpersonal violence problems, and physical health problems). CGPT was able to understand 91% of the questions and answered 21 out of 23 questions with clear and appropriate language. The result is significantly superior to the performance provided by Amazon Alexa, Apple Siri, Google Assistant, Microsoft's Cortana, and Samsung's Bixby in cases where these AI systems have been used as consultants for health problems [3].

Another group of researchers evaluated the diagnostic capabilities of CGPT by proposing as queries the clinical cases proposed by the New England Journal of Medicine from January 2021 to December 2022, obviously omitting the diagnostic assumptions made therein (Not only the level of accuracy of the data is well known but also the complexity of the NEJM clinical cases). Also, in this research, the result was quite satisfactory for CGPT; in 39% of the cases, the first diagnosis made was found to be correct, and in 64% of the cases the correct diagnosis was still listed in the differential diagnostic hypotheses made by CGPT.

The researchers compared the performance provided by CGPT with that of various automated digital differential diagnostic systems; the results were comparable with the best among these diagnostic systems, some of which were, however, superior in the correctness and accuracy of the diagnoses [4].

On the basis of the data available so far, CGPT seems very promising as an assistant in diagnosis, as communicative support in the doctor-patient relationship, and as an aid of the patient in adherence to therapy and in all rehabilitation pathways with particular regard to neurological and orthopedic rehabilitation.

In summary, we can say that CGPT and the upcoming advanced AI systems available to the public have opened a new era: physicians and health professionals will either use them occasionally by drawing news here and there or appreciate them for their great potential by constantly evaluating their possible limitations and errors.

In any case, it is critical to remember that CGPT is an artificial intelligence model and cannot replace the expertise and advice of a qualified physician. Physicians should use it as a support tool and integrate it with their clinical expertise to make informed decisions about patient care.

In particular, it is imperative a regulatory oversight to assume medical professionals and patients can use CGPT and other large language models without causing harm or compromising their data or privacy [5].

References

[1] Galimberti U. Nuovo Dizionario di Psicologia. Milano: Feltrinelli Editore; 2018.
[2] Ayers JW, et al. Comparing physician and artificial intelligence chatbot responses to patient questions posted to a public social media forum. JAMA Intern Med 2023;. Available from: https://doi.org/10.1001/jamainternmed.2023.1838.
[3] Ayers JW, et al. Evaluating artificial intelligence responses to public health questions. JAMA Netw Open 2023;6(6):e2317517. Available from: https://doi.org/10.1001/jamanetworkopen.2023.17517.
[4] Kanjee Z, Crowe B, Rodman A. Accuracy of a generative artificial intelligence model in a complex diagnostic challenge. Res Lett JAMA 2023;. Available from: https://doi.org/10.1001/jama.2023.8288 Published online June 15.
[5] Meskò B, Topol E. The imperative for regulatory oversight of large language models (or generative AI) in health care. Npj Digital Med 2023;6:120. Available from: https://doi.org/10.1038/s41746-023-00873-0.

Glossary

Information and communication technologies or ICT technologies concerning the treatment and transformation of data that are transmitted in information processes (integrated telecommunications systems, computers, audio-video technologies and related software)

Big data collection of data so huge in volume, speed, and variety that it requires specific and complex technologies capable of processing analysis and from which to obtain useful information

Secondary data data generated for purposes other than research

Applications (apps) application software designed for mobile devices (smartphones, tablets, and smartwatches)

Mobile Health healthcare model created through the use of mobile devices and multichannel technologies, such as cell phones, smartphones, wearable devices, handheld *personal digital assistants*, and other wireless devices

Cloud computing series of *hardware* and *software* technologies that allow data to be processed and stored over the Internet

Point-of-care testing medical examinations at (or near) the patient's place of care

Digital divide the gap between those who have effective access to the Internet and those who are excluded, partially or totally

Internet of Things *a network of objects connected to the Internet capable of collecting, recording, analyzing, and sharing data using sensors and other technologies*

Algorithm *(from medieval Latin algorithmus, derived from the name of the Arab mathematician Muhammad Ibn Musa Al-Khwarizmi)* series of precise instructions in mathematical language to be carried out to obtain the solution of a given problem, for example, to find associations, identify trends, and extract the dynamics of the financial markets

Restricted or weak artificial intelligence machine capable of simulating/surpassing human intelligence in a very specific task

General or strong artificial intelligence machine capable of emulating and even surpassing human intelligence in complex tasks

Technological singularity idea that, at some point, machines will be intelligent enough to program and improve themselves, to the point of becoming independent

***Machine learning* or machine learning** branch of AI in which computers provide answers to problems presented, in an adaptive, iterative, and independent way, "learning" from the data introduced to identify precise relationships in the observed data, for classification and prediction purposes, without rules and explicit preprogrammed models

Reinforcement learning the system's ability to continually refine its ability through a series of reinforcements (rewards or punishments)

Unsupervised learning the system's ability to learn without explicit feedback. The algorithm must be able to infer a rule that groups the cases that arise by inferring the characteristics from the data itself.

Supervised learning ability of the system to learn through externally provided "examples," categorized a priori, until the algorithm reaches an acceptable level of performance

Antagonistic learning principle used by the AlphaGo Zero system, in which two neural networks face each other, each of which to overwhelm the other, initially exchanging random "hits" and then refining, game after game, their strategies to possess, in a short time, a profound mastery of the game without being aware of historical games

Neural networks information processing systems that try to simulate the functioning of biological nervous systems. In summary, they are made up of hundreds of layers of interconnected "neurons" (actually conducting material), which acquire information, process it and propose new content to the next level, which will resume the process by transmitting the data to a further level, and so on until you reach the desired goal. Each link has a "weight" that measures how important the link between two particular "neurons" is

Deep learningmachine learning method, which can be translated as deep learning, based on neural networks with a very large number of levels or layers

Data augmentation add different typologies to the data submitted to the system to teach you even small variations on the theme, for example, in the case of a cystic lesion on ultrasound, provide images with internal spots to allow the recognition of a "not perfect" cyst

Natural language processing processing of *an unstructured, spontaneous text in a structured form and therefore interpretable by a machine for automated data processing*

Cognitive computing system capable of improving the human-machine interface through the implementation of natural language

BOX 1 Artificial intelligence

Know it to use it in the best ways

What are the characteristics of human intelligence so far reproduced (and sometimes exceeded) in artificial intelligence systems?

The ability to reason, or to produce, a symbolic representation of the world that allows us to understand the events examined and to formulate projects that will be realized through planned actions.

The ability to solve problems through logical processes

The ability to learn and to modify the external environment through a modification of one's actions such that they allow to reach the set objectives [1].

What features are not reproduced in any artificial intelligence system?

Intuition, which deep learning sometimes seems to manifest by proposing unexpected and creative solutions, which are, however, based on thousands of very fast probabilistic assessments among which it selects the most suitable for the objectives

Emotional intelligence, or the ability to recognize, use, understand, and consciously manage one's own and others' emotions.

Creativity, which is probably the result of the two human faculties mentioned earlier, which is the basis of artistic creation and of all the great scientific discoveries

Artificial intelligence systems have immense capabilities but, luckily for us, they only act within the universe of the algorithm that generated them [2].

Processes used in artificial intelligence systems

Induction is a logical procedure by which, starting from the observation of particular cases, one arrives at a conclusion valid for multiple cases, ideally for all cases similar to those observed [2].

Deduction

Deduction is the logical procedure by which, starting from one or more premises, one arrives at a certain conclusion, valid in reference to those precise premises [3].

Analogy

Analogy is a knowledge tool that refers to the similarity between situations and processes that have certain common characteristics: it is used for an unknown situation but similar to a known one, a classification and possibly a solution that has already proved effective. (In the philosophical field, analogy has been used as an instrument of knowledge since Plato and Aristotle; its cognitive limits have been well highlighted in particular by Kant) [3]

However, it still remains a tool used profitably by many artificial intelligence systems, for example, for the psychological profiling of user.

The Joker of Statistics: Bayes

The use of the enormous calculation speed of computers to make probabilistic estimates based on the Bayes Theorem has opened new perspectives to artificial intelligence: as is well known, Bayes' theorem allows to estimate conditional probabilities with increasingly accurate levels of precision, that is, probability that an event B will occur following the occurrence of a known event A: on the basis of the calculation, the most probable solution is chosen, often tested in virtual environments and corrected one or more times.

Evolutionary adaptation software and genetic algorithms

They are based on this versatile theorem and reproduce in the research process, on the one hand, the random genetic mutations, sometimes very useful, on the other hand, the mechanisms of natural selection: in other words, many different solutions are tried which are gradually eliminated by the selection mechanisms or software: if we remember that modern powerful microprocessors carry out thousands of calculations per minute, we understand that computers are not intelligent but are only formidable workers who sooner or later "get it right" by mathematical laws [4].

Neural networks and deep learning

After decades of disappointing experiments at the end of the 1990s, it was decided to "imitate" (analogy) the human brain: electrical circuits were designed that reproduced in part the complex neuronal interactions and tested them: it was a success.

Neural networks are generally made up of several successive layers, each made up of circuits that interact within each layer; the first activated layer transmits to the second layer a solution which is tested and corrected.

The process is repeated several times in several layers and the results provided are the result of random combinations of data that are repeatedly filtered by eliminating the combinations deemed useless, impossible, or inconsistent until reaching an optimal result.

References

[1] Pareschi R., Dalla Palma S., Artificial intelligence. A journey among the thinking machines that will change our future. Hachette issues ed. Milan; 2019.
[2] Domingos P., The definitive algorithm. Bollati Boringhieri Ed. Turin; 2018
[3] Treccani Italian Encyclopedia. <http://www.treccani.it/encyclopedia/ricerca>.
[4] Kurzweil A., How to create a mind. The Secrets of Human Thought. Apogeo Ed. Milan; 2013.

Index

Note: Page numbers followed by "*b*" and "*t*" refer to boxes and tables, respectively.

D

N

O

P

W

X

Printed in the United States
by Baker & Taylor Publisher Services